Reference Systems and Inertia

The Nature of Space

Reference Systems and Inertia

THE NATURE OF SPACE

Beryl E. Clotfelter

THE IOWA STATE UNIVERSITY PRESS / AMES, IOWA

To My Wife, Mary Lou

BERYL E. CLOTFELTER has been Professor of Physics at Grinnell College since 1963. He has also taught at the University of Idaho and Oklahoma Baptist University and has had research experience in industry. He holds a B.S. degree from Oklahoma Baptist University, the M.S. and Ph.D. from the University of Oklahoma, and also did graduate work in physics at Ohio State University. More recently he spent a year at Princeton University studying general relativity and cosmology as a National Science Foundation Fellow. His previous publications have been papers in *Physical Review, Journal of Applied Physics, Journal of Chemical Physics,* and *American Journal of Physics.*

Composed and printed by
The Iowa State University Press

First Edition, 1970

International Standard Book Number: 0-8138-1325-5
Library of Congress Catalog Card Number: 73-1144796

Contents

Preface

SEVERAL TIMES in the past, discoveries about the physical world have produced radical changes in man's concept of the universe. In the course of these our universe has changed from a small, geocentric universe to one that is unimaginably vast, in which the earth is a speck; from a static world to a changing, expanding world; from a world in which Newton's mechanics was the ultimate in physical theory to one in which relativity and quantum mechanics are the everyday tools of the physicist. Perhaps similar changes in our view of the physical world will never come again, but if they should, the most likely places for the critical discoveries to be made are in the area of elementary particles and in the area of relativity and cosmology. In the one the fundamental question is "What is the ultimate structure of the world on the very smallest scale?" and in the other it is "What is the ultimate structure of the world on the very largest scale?" Neither question is new, and neither question has been answered, but in both areas answers to ancient problems appear tantalizingly close. And since these questions deal with the most fundamental questions of physical science, they are among the most interesting for students.

The goal of this book is to make available to students who have a minimal background in physics an introduction to current work and modern thinking on one aspect of the nature of space and the related problem of the origin of inertia; these are subsidiary questions to the ultimate question of the large-scale structure of the universe. The mathematics has been held to a minimum in order that the explanations will not appear forbidding to students who may have had only enough for a beginning course in college physics. References are given to original papers so that those who wish more detailed descriptions of experimental work can find them, but no attempt has been made to give a complete set of references to all related work. The emphasis is upon experimental work which is intended to help answer fundamental questions; in order to make the experiments intelligible a considerable amount of discussion of theory has been included, of course, but the orientation of the book is toward experimental physics rather than toward philosophy. At the same time the questions treated are partially within the realm of philosophy, and it is not inconceivable that students of the philosophy of science might find the discussion of interest.

Reference Systems and Inertia

The Nature of Space

1
Frontiers of Mechanics

Some of the most interesting and challenging problems in physics are to be found in the area of mechanics. This statement is true in the second half of the twentieth century even as it was in the second half of the seventeenth century, for contrary to the impression a beginning student in physics may have, mechanics is not a closed subject. Mechanics still contains "frontiers of knowledge," even though most of the methods and facts presented in courses in classical mechanics were known in finished form a century ago and the special theory of relativity which is often included in those courses was presented in almost complete form over half a century ago. Problems of fundamental concepts with which Newton wrestled are still with us, and some of the questions which he raised do not yet have satisfactory answers.

In this book we shall consider two of the problems which Newton left for us, both fundamental to an understanding of classical mechanics. We shall see that some of the definitions and laws given by elementary textbooks and learned by students conceal more than they reveal, and that mechanics can be considered a closed subject only if one restricts his attention to certain types of manipulations; if he looks more deeply, he finds it is a live, very modern subject which is receiving intensive attention now from both experimental and theoretical physicists.

The problems we shall consider are those of the definition of an appropriate reference system for the application of Newtonian mechanics (a Galilean reference system) and the problem of the nature of mass. Neither problem has been "solved," and the discussion will be more an introduction to an area of modern research than a presentation of neatly packaged answers.

The problem of choice of reference system raises profound questions about the nature of space itself, but the initial problem can be stated very simply thus: Newton's laws of motion are statements about acceleration. But acceleration must be measured relative to something. With respect to what should one measure acceleration to obtain numbers to put into the equations expressing Newton's laws?

A reference system may be visualized as three mutually orthogonal rods which are marked off in feet or meters. If a body is moving relative to this rigid system, its position can be noted at successive times and its velocity and acceleration computed. (We will restrict our attention for the moment to small distances and low velocities so that problems of relativity can be ignored.) One can imagine many reference frames, moving with respect to one another, in rotation, and in general very arbitrarily chosen. One might be fixed to the earth at one location, another fixed to the earth at a different point, one fixed with respect to the sun, another attached to the airplane passing overhead at a particular moment, and so forth. Clearly a particular object may be motionless in some such reference systems, moving with constant velocity with respect to others, and it may have accelerations in still others. Our important question is, then, not How can we measure acceleration? but How shall we choose a reference system in which Newton's laws of motion will be valid? We shall use the first of Newton's laws as the test and try to find a reference system in which a body acted on by no external force will remain at rest or continue in motion at constant speed in a straight line. If we can find such a system, we shall call it a Galilean reference system; and we can be sure that in that system the other two laws of motion will be valid. The paramount question, therefore, is, Do Galilean reference systems exist, and if they exist, how do we find and recognize them?

In most of the problems considered in introductory physics courses, and indeed in most practical applications of physics in ordinary experience, one measures motion and acceleration relative to a point or a line on the surface of the earth. A block slides across a table with increasing speed; the speed and acceleration are measured relative to the table, which is fixed on the surface of the earth. Or an automobile accelerates on a roadway; the motion is considered relative to the road.

Clearly, however, such reference systems do not have universal applicability; that is, they are not Galilean reference systems. For example, as the earth spins and moves around the sun, distant stars alternately approach and recede from my table in the laboratory or the stretch of road in front of my house, and I cannot be so egotistical as to think that because the stars appear to me, from my reference system, to be accelerated toward or away from me that therefore forces (of magnitude:

mass of star × acceleration of star) are acting on them. And in fact, both the rotation of the Foucault pendulum and the drift westward or eastward of winds and ocean currents in response to the Coriolis force give evidence that the reference systems attached to the surface of the earth are not entirely correct for the application of Newton's laws of motion.

But, the reader may be saying, perhaps these problems with a reference system fixed on the surface of the earth arise because the earth is rotating on its axis. If we measured accelerations relative to a system fixed at the center of the earth, we would eliminate the difficulties. It is true that such a change would reduce the problems, but let us look at the matter in this way: According to Newton's first law a body at rest, acted upon by no forces, will remain at rest. Let us imagine a body far enough out in space that the attraction of the sun and earth for it can be neglected for a short time, and let us suppose that the body is released at some time motionless with respect to the center of the earth. If the coordinate system fixed at the center of the earth is a proper one for the application of Newton's laws, the body should maintain its position relative to the earth. But does anyone suppose that it will do so? The earth is moving around the sun at approximately 18½ miles per second, and it is traveling in a curved path. Surely the earth will "run away from" the object suspended in space so that the object will appear to accelerate away from the earth. But since no forces are observed to act upon it to produce the apparent acceleration, we must conclude that the reference system chosen was improper.

The same argument can be applied again if we

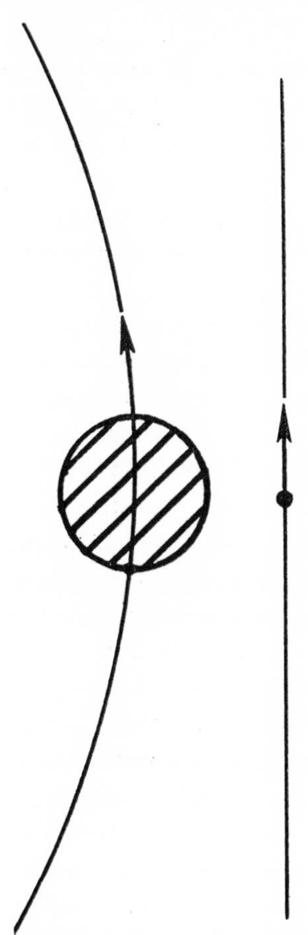

FIG. 1.1. An object at rest with respect to the center of the earth at one instant has a velocity equal to the earth's orbital velocity at that instant. As time goes by, the object continues at constant velocity, on a straight line, but the earth curves away from it.

choose a reference system fixed in the sun, for the sun is moving around the center of the galaxy in a curved path which is approximately a circle about 30,000 light-years in radius, and its speed along this path is approximately 200 mi/sec. Likewise, the center of the galaxy is not entirely satisfactory, for it appears that our galaxy is one of a group of galaxies which probably are rotating about a common center. Viewed from this perspective, the remarkable thing about the table or road on the surface of the earth is not that it is not a perfect reference system, but that it is usable at all as an approximation to the ideal system. As we consider what would be the ideal system, it is hoped we will be able to understand why what appears to be a poor approximation actually serves so well.

Newton solved the entire problem of choice of reference systems neatly by announcing that there exists an absolute space, motion relative to that space is absolute motion, and the accelerations described in his laws of motion are to be measured with reference to absolute space. This is a lovely solution of the problem, and if we could find experimental evidence for the existence of absolute space, we should have no more worries. But the fact is that all attempts to detect it have been in vain. We must raise the question, then, as to whether absolute space exists, and what significance we can attach to discussions of something which seems to be beyond our detecting. At the very least it is disconcerting to discover that the accelerations so casually discussed in beginning physics must be measured relative to a reference system no one has yet found in order that Newton's first and second laws shall be correct!

Turning to the second question to be considered in this book, we can pose the problem of the origin of inertia in this way: Newton's laws, and indeed our own experience, tell us that when a massive body is accelerated, it tends to resist that acceleration. We call the property of matter which causes such resistance to acceleration "inertia." The interesting question is,

Is inertia an intrinsic property of matter or is it the result of the interaction of matter with matter, somewhat like gravitation is the interaction of one body with others? The meaning of this question will become clearer as we continue. Newton thought that inertia is an intrinsic property; but it is easy to call this view into question, even if it is not easy to find a satisfactory alternative. In fact, one of the best examples to cast doubt on the idea that inertia is purely intrinsic is an example Newton himself used to "prove" something else.

Newton suggested that we think of a bucket of water suspended by a rope and imagine the bucket turned around and around until the rope is twisted. If the bucket is released, the rope unwinds. At first the water in the bucket does not rotate and its surface is level, but soon it begins to turn with the bucket and the surface becomes concave upward. If after a time the bucket is stopped, the water continues rotating and the surface remains curved, until finally the water comes to rest, at which time the surface is again plane. Newton's argument was that we can detect absolute rotation—rotation relative to absolute space—by looking for the curvature of the surface of the water. Ernst Mach, a German physicist of the late nineteenth century, however, interpreted the results differently. The only rotation we know about, Mach said, is rotation relative to the other objects in the universe—the earth, the sun, the stars, and the galaxies. If all these were removed from the universe, "rotation" would have no meaning; and we must presume that the surface of the water in the bucket would always be flat. Since the rotation relative to other masses external to the system we are considering (bucket and water) is the important thing, our analysis should use only that relative motion; the description of a stationary universe and rotating water or stationary water and a rotating universe should be the same, for the two cases are experimentally indistinguishable. In other words, a man riding on the bucket and observing the water surface assume a curvature should be able to explain the curvature by noting that the universe apart from the bucket has begun to rotate around him. It is a defect of our formulation of the laws of motion, said Mach, that they treat the rotating mass of the water differently from the other masses, so that we are led to suppose that we should get one result if the water rotates and we imagine the stars to remain fixed but get another result if we let the water remain fixed and imagine the stars to revolve about the bucket. (This sort of distinction makes sense if there is in fact an absolute space, for then one can say that the water actually is moving and the stars are standing still, but if there is no such thing as absolute space, one can as easily imagine the universe to spin as the water. Well, *almost* as easily!) But now let us think of the case in which the water sits still and the remainder of the universe revolves around it. The surface of the water must become concave. But why should it? Clearly if the water moves in response to the rotation of the stars and galaxies about it, those bodies must have some effect on the water. And if the water is rotating, it must be some interaction between the water and other matter in the universe which causes it to seek the outer part of the bucket and thereby causes curvature of the surface. But the tendency of the water to recede from the axis of rotation is one manifestation of the property we call "inertia," and hence we must conclude that inertia seems to be a result of interaction of matter and matter.

But if in the case of rotation the effects we usually term inertial are the result of an interaction with distant matter, may not the effects of inertia in ordinary, linear acceleration also be the result of some interaction with distant matter? Let us consider a gedanken experiment like this: Suppose that all the matter in the universe has been removed except one massive body—a brick for example. We shall need two specks of glowing material of negligible mass to provide a reference system. Let us imagine the brick placed between the glowing specks and at rest with respect to them; the specks also are at constant distance from one another. If while the specks maintain their distance from one another, the brick suddenly begins to accelerate toward one and away from the other, Newton's first and second laws tell us that a force must be acting upon the mass. (Don't worry about the origin of the force in such an empty universe; that is one of those nit-picking questions that confuse a good argument!) The proper question is, Would the force required to produce a certain acceleration in the empty universe be the same force required in our normal universe? If we could believe that there is an absolute space, so that we could say that the brick is absolutely accelerated, it might be easy to answer the question Yes. But if there is no absolute space and we have no way of measuring acceleration except by comparison with the two glowing specks, the question becomes more difficult. If inertia is an intrinsic property of matter, the answer is Maybe, but if inertia is a result of the interaction of matter with other matter, the answer is No. (Even if inertia is an intrinsic property, the answer is Maybe instead of Yes because of uncertainty about the reference system. How is one to know that if the brick suddenly lunges toward one speck and away from the other, it is because something has happened to the brick instead of to the specks? The questions of reference system and origin of inertia are not entirely separate matters.)

If in our empty universe we place a bucket of water and four glowing specks arranged in a square about the bucket, we can ask questions about the result of rotations. Suppose we spin the bucket until the water is also spinning relative to the four specks. Does the water surface curve upward at the outer edges? Suppose that the bucket stands still and the four specks are caused to rotate around it. (The bucket "stands still" in the sense that we do nothing to it; we take hold of the specks and move them.) Does the surface of the water change? Or will the surface of the water remain plane regardless of the relative motion of the bucket and the specks because there is not enough matter in the universe to cause it to curve? Newton could answer that if the bucket were rotating with respect to absolute space the water would be curved, but few modern physicists would be as certain as he about the correct answer.

Some physicists may object that this is all very interesting, but is philosophy rather than physics and is of little concern to the working physicist. It is true that objections to Newton's absolute space and absolute motion were first raised by philosophers—Bishop Berkeley was probably the first—and it is also true that such questions are of great interest to many philosophers today. But the problem is also one of physics. Without going into arguments about the distinction between philosophy and physics, we may suggest that a workable criterion for problems which are legitimate physics is the possibility of seeking or finding experimental answers. Physics is an experimental science, and all its propositions and laws should have some intimate connection with experiment. The reason for the inclusion of questions about absolute space and the origin of inertia in physics rather than exclusively in the philosophy of science is that experiments can be devised to shed some light on the questions, even if the unambiguous answer has not yet been obtained. The interest in experimental attacks on these questions has greatly increased in the 1960s; and experiments of remarkable precision have been performed, which have some bearing on the questions we have raised.

In the chapters to follow we shall consider the problems in somewhat more detail and then shall discuss the experiments which are relevant to each. Finally, we shall attempt to indicate where we are with regard to some definitive answers and to suggest the trend of modern research. And so—first to reference systems and then to inertia.

2
Reference Systems

A REFERENCE SYSTEM is a device for locating points in space, and the kinds of reference systems we imagine are closely related to our ideas of the nature of space itself. The nature of space had been debated by philosophers for a very long time before Newton lived; but until Newton developed his system of mechanics, the discussion of space had had little to do with science. Two questions may be asked about the nature of space which may be debated almost endlessly: Is space infinitely divisible? and, Is there an absolute space? The first question will not be considered in this book, for although there have been speculations on it within the past 50 years, no useful evidence to permit an answer is at hand. The development of the quantum theory and the realization that electric charge, mass, and energy are all quantized has naturally led to suggestions that perhaps space is also quantized and a spatial interval cannot be made arbitrarily small; but no experimental evidence or convincing theoretical arguments have been found for such quantization.

A great deal more can be said about the second question—the absolute nature of space—and with this we shall concern ourselves. As discussed by philosophers, the question tended to focus on the relation of space to matter: could space exist if there were no matter in it, or does the concept of space require the existence of reference material by which distances can be defined? Is our concept of space an extrapolation from the definite idea of distances between physical objects, or does space refer to something with an independent existence in which physical objects are embedded? As long as the questions are left in this form, answers are chiefly a matter of opinion. Perhaps psychology could give some insight into the way such concepts are formed, but it is difficult to formulate experiments which might suggest answers to the fundamental questions. The only space we know does have matter in it, and what might exist if there were no matter is beyond the range of our experimental test.

The questions about the nature of space take on a somewhat different character in mechanics, however, for the attention is shifted from existence and ultimate nature to dynamic behavior. Space enters into the theory through the description of motion, and the questions about the nature of space can be transformed into questions concerning velocities and accelerations and other entities such as force, which are related to them. One can hope that the relevant questions concerning the motion of physical objects might be susceptible of experimental answers, and thus the more inaccessible properties of space might be deduced.

Newton's laws of motion, the foundation of classical mechanics, demand an absolute space; that is, if they are to be logically complete, there must exist a space in which things are absolutely at rest or are absolutely moving. Of this there was no doubt in Newton's mind, as the following quotations from the first section (Definitions) of the *Principia* reveal:

> Absolute space, in its own nature, without relation to anything external, remains always similar and immovable. Relative space is some movable dimension or measure of the absolute spaces. . . .
>
> Absolute motion is the translation of a body from one absolute place into another; and relative motion, the translation from one relative place into another. Thus in a ship under sail, the relative place of a body is that part of the ship which the body possesses; or that part of the cavity which the body fills, and which therefore moves together with the ship: and relative rest is the continuance of the body in the same part of the ship or of its cavity. But real, absolute rest, is the continuance of the body in the same part of that immovable space, in which the ship itself, its cavity, and all that it contains, is moved. . . .
>
> The causes by which true and relative motions are distinguished, one from the other, are the forces impressed upon bodies to generate motion. True motion is neither generated nor altered, but by some force impressed upon the body moved; but relative motion may be generated or altered without any force impressed upon the body. For it is sufficient only to impress some force on other bodies with which the former is compared, that by their giving way, that relation may be changed, in which the relative rest or motion of this other body did consist.[1]

By introducing the concept of absolute space, Newton solved one problem, for he made possible a very successful system of mechanics. In fact, the success of his

1. I. Newton, *Mathematical principles of natural philosophy*, trans. Andrew Motte, rev. Florian Cajori (Berkeley and Los Angeles: Univ. Calif. Press, 1962). (Reprinted by permission of The Regents of the University of California.)

system during the century following the publication of the *Principia* was so great that objections to it obtained slight hearing. Who could quarrel with a formulation of mechanics which could explain the motion of projectiles, planets, comets, and all other bodies to which it was applied and which seemed capable of indefinite extension and refinement? But by introducing the concept of absolute space, Newton raised another problem—namely, how does one detect this absolute space, and how does one decide when or if he is absolutely at rest? As physics has become more concerned with eliminating those concepts which cannot be given meanings relevant to experiment, absolute space has posed a particular problem, for on the one hand Newtonian mechanics is a complete and satisfactory description of the world only if it is included, and on the other hand, it seems to have no relevance to experiment. In fact, such experiments as those performed by Michelson and Morley or by Cedarholm and his collaborators seem to indicate that no experiment can detect absolute space or absolute motion.

Two questions are likely to arise in the mind of the reader at this point: (1) Was Newton right in thinking that his theory demands absolute space? and (2) Does not the theory of relativity satisfactorily answer all these questions? Let us consider the first question now and defer the second for a short time.

The notion of absolute space is likely to seem superfluous because of the way in which Newtonian mechanics is usually related to inertial reference frames (Galilean reference frames). By definition an inertial reference frame is a reference system in which Newton's first law is true, that is, one in which a body acted upon by no forces will remain at rest or continue in motion at constant velocity. We may visualize such a reference system as being marked off by three mutually perpendicular lines, or we may think of the particular part of it to which our attention is directed as comprising a very large room. It is important to realize that the reference system could be extended to include an unlimited amount of space, but that we focus our attention upon a rather small part of it. If a golf ball placed at an arbitrary point in the space and released at rest relative to the reference system remains at rest as long as no forces act on it, we may be sure that the reference system is truly an inertial system. That is quite clear and simple. But how do we determine that a force acts on our test object, golf ball, or something else? The criterion for recognizing the presence of an unbalanced force is the production of acceleration! It appears then that as stated the first law borders on the vacuous, for it tells us that when viewed in a proper reference system a body acted on by no forces exhibits no acceleration; and we determine that no forces are acting by observing the lack of acceleration. In practice, however, this circularity is usually not a problem because we assume that an uncharged body, not affected by the gravitational field of any other body, not pulled by strings or pushed by rods, and not affected by anything else we commonly associate with "forces" is free from force. In other words, if we can see no agency to produce a force on the body, we assume that none exists. The requirement that the body not be subject to gravitational forces poses a real problem, for where in the universe is one to get away from gravity? Obviously the answer is Nowhere. Does this mean then that no inertial reference system exists anywhere? Not necessarily, although it probably does mean that there is no possibility of finding one inertial reference system which will extend throughout all space or even will extend for a vast distance. Instead we are forced to look for reference systems which are inertial throughout a limited region. Two circumstances make possible the location of local reference systems which are inertial. One is that the requirement is not that the gravitational forces be zero, but that the net force produced by gravity should be zero. Because of the apparent homogeneity and isotropy of our universe when viewed on a sufficiently large scale, one may expect that at a point far out in intergalactic space the gravitational forces on a test body would be so nearly the same from all directions that their vector sum would be zero. The second circumstance is that a reference system falling freely in a gravitational field acts as if it were gravity-free, and such a system is a good approximation to the ideal Galilean system over short distances. In an elevator which is falling freely or in a satellite in orbit around the earth in "free fall," all objects respond to gravity in the same way; hence it seems to produce no effects. An apple released in space a foot before a man's face remains there—the test to indicate that the elevator or satellite is at least a good approximation to an inertial reference system.

The surface of the earth clearly is not such an inertial system; how does it happen then that we can treat laboratory tables and stretches of highway as if they were proper reference systems? The answer is that for many purposes we apply a vertical force of support which effectively balances the gravitational force on our test objects, so that the bodies act as if they were free of gravity. The rotation of the earth on its axis and its motion around the sun and about the galactic center are defects of the reference frame which cannot be eliminated, but the centripetal accelerations resulting from those rotations are so low that for many purposes they can be neglected. For more precise work, a reference system can be defined with reference to the "fixed stars" which seems to be a quite satisfactory inertial system.

If we can find one inertial system, any number of others can be found, for any system moving at constant velocity (constant speed in a straight line) relative to the first system is also an inertial frame. If an object is at rest in the original reference system, it moves at constant velocity in each of the others. If it has an acceleration in one system, it has exactly the same acceleration in each of the other systems. In any one of these inertial systems Newton's laws of motion are valid.

It may seem, then, that no reference needs to be made to absolute space. But let us consider this situation. Suppose that we are observing a body (or many bodies) either at rest or moving in a straight line relative to some coordinate system. If we have no reason to suppose that forces are acting on the bodies, we will conclude that the reference system in which we are observing them is an inertial frame, for that is the only criterion we have for choosing inertial systems. Now suppose we suddenly observe that the bodies begin to accelerate relative to the reference system. What conclusion are we to draw? One possibility is that a force is now acting on the bodies such that their accelerations are given by the familiar $F = ma$. An equally valid conclusion, however, is that no force is acting on the bodies but that the reference system (which must involve material bodies having mass) is being acted upon by a force so that it is accelerating. If, as is actually the fact, relative motion is all we can observe, the observed facts in this case can be explained equally well by either explanation. From the standpoint of Newton's laws, however, the explanations are not equivalent. We must determine in some way whether the reference system is still an inertial frame in order to know whether a force is acting on the bodies. The only way such a determination can be made (if all the bodies we can observe are showing the same acceleration) is to compare with another inertial system; but that system can be established as inertial only by comparison with a third, and this sort of argument can be continued indefinitely with the crucial comparison always being with the next system. We conclude, therefore, that if we wish to be able to give definite meaning to the statements that bodies will show accelerations only when forces act on *them* in distinction to the apparent accelerations relative to neighboring bodies observed when forces act on those neighbors, we must have an absolute determination of an inertial system. In other words, there must be some system which is absolutely at rest, and material bodies will have accelerations relative to it only when actual forces act on them. In an inertial reference system the laws of motion are valid; but unless we admit the notion of absolute space, we can define inertial systems only as those systems in which the laws are valid. Thus we are caught in a circular argument; in practice we either assume that a system which has once been shown to be inertial (or a reasonable approximation) remains inertial and attribute changes of bodies within the system to forces acting on them, or we are able to test for the inertial character with some bodies while at the same time we are applying forces to other bodies and observing their accelerations. If we find few problems in knowing how to apply the laws of motion to real situations, this may be more because we usually are satisfied with a fair approximation to the behavior predicted by the laws and we commonly work with simple and obvious applications of force—a stretched rope, a charged body in an electrical field, a jet of hot gas, and the like——than because of the logical consistency of our analysis. Logical analysis forces us to the conclusion, however, that Newton was right, and the existence of absolute space is a necessity if his system of mechanics is to be consistent.

Undoubtedly physicists of an earlier time had thought about trying to measure the velocity of the earth relative to absolute space, but not until the latter part of the nineteenth century were experimental methods developed of such sensitivity that the attempt became feasible. Beginning in about 1880 and continuing down to the present time, a series of experiments of increasing sensitivity have been carried out in attempts to detect motion relative to absolute space; the most important of these experiments will be discussed here.

These experiments are all referred to loosely as "ether-drift experiments," referring to the reasoning usually associated with the first and most famous, the Michelson-Morley experiment. To understand this reasoning, we must consider the ether and the role it was assumed to play.

When in the early nineteenth century light was recognized to be a wave motion, the necessity for it to be a wave in something inspired the invention of the ether. The ether was a tenuous material which pervaded all space, which impeded the motion of the planets negligibly, but which was the medium through which light traveled. Later, about the middle of the nineteenth century, when James Clerk Maxwell developed the theory of electricity and magnetism and found that the theory led naturally to the prediction of electromagnetic waves which should travel at about the speed at which light was known to travel, it was recognized that light probably consists of traveling electromagnetic fields, and the obvious conclusion was drawn that electric and magnetic fields also represent a disturbance of the ether. Maxwell's theory of the fields predicted precisely what the speed of propagation of the waves should be in free space, but it was not explicit about the reference system in which the velocity should be measured. This seemed to pose no problem, however, for the natural way to measure the speed of any wave is with reference to the medium through which it is passing. For example, the speed of a sound wave in air is given by $\sqrt{\gamma P/\rho}$, where γ is the ratio of the specific heat at constant pressure to the specific heat at constant volume, P is the pressure, and ρ is the density. The formula as stated gives no indication of the reference system with respect to which the speed is to be measured; but if we should look at the derivation of this formula, it would become clear that the thing being calculated is speed relative to the air itself. Analogously, it was assumed that the speed of Maxwell's electromagnetic waves, now identified with visible light, was to be measured with respect to the ether in which they traveled. But if the ether pervades all space, what is more natural than to assume that it is at rest in absolute space? Thus, if one could measure the speed of the earth relative to the ether, he should have a measure of its speed relative to the absolute space required by Newton's mechanics. And if absolute space seemed hard to "get hold of" experimentally, the same was not true of the ether, for motion through it should produce effects which could be detected and measured in the laboratory quite unambiguously—if not quite easily.

During the nineteenth century various attempts were made to detect the motion of the earth through the ether—the so-called ether drift or ether wind—but prior to the time Albert Michelson took up the problem, all the measurements were made looking for effects proportional to (v/c), where v is the velocity through the ether and c is the velocity of light. The attempts to find effects which could be attributed to the motion gave uniformly negative results.

About 1880 Michelson became interested in the problem; and specifically for the purpose of making more sensitive measurements than had been made previously, he invented the interferometer which bears his name. He was inspired in this by the fact that Maxwell's theory of electromagnetism predicted that although effects based on it proportional to (v/c) should not be detectable, those proportional to $(v/c)^2$ should be. Since the earth has an orbital speed of the order of 18 mi/sec, and the speed of light is approximately 186,000 mi/sec, the square of the ratio of the two numbers is very small; consequently, a sensitive measurement is required. This, however, the interferometer could provide.

Michelson had an interferometer made for this

purpose and first tried the experiment in Europe in 1881.[2] The results were negative—no motion through the ether was found. He was not satisfied with the experiment, in part because the rotation of the apparatus which was a necessary part of the procedure introduced strains which reduced the sensitivity; therefore, six years later he repeated it with greatly improved equipment and with a collaborator—Edward Morley. This was the famous Michelson-Morley experiment, carried out at Cleveland in the summer of 1887.[3] Subsequently Michelson and others repeated the experiment many times in essentially the original form.

The Michelson interferometer consists of 1 partially silvered mirror; 2 plane, totally reflecting mirrors; and a compensating plate made of the same glass as the partially silvered mirror. It may also include a telescope to aid in viewing the fringe pattern. In the form used by Michelson and Morley the instrument was mounted on a sandstone block 5 feet square and 1 foot thick, and other mirrors were added to make the total length of light travel in each arm of the interferometer 1,100 cm. The experiment consisted of rotating the equipment through 90° and looking for a shift in the pattern of fringes seen through the telescope; it was this rotation which had introduced problems into the work done in Europe in 1881. To permit rotation without strains, Michelson and Morley supported their stone block on a wooden annular ring which floated in mercury contained in an annular cast-iron trough. A pin at the center held the entire apparatus centered while rotation was begun; then the pin was disengaged.

2. R. S. Shankland, *Am. J. Phys.* 32 (1964):16.
3. A. A. Michelson and E. W. Morley, *Am. J. Sci.* 34c (1887):333.

When the block had been set rotating at a rate of 6 minutes per revolution, it would continue to turn for hours; and the rate of rotation was so low that observations were made without stopping or touching the apparatus.

Now we shall look at the theory of the Michelson-Morley experiment and make some approximate calcu-

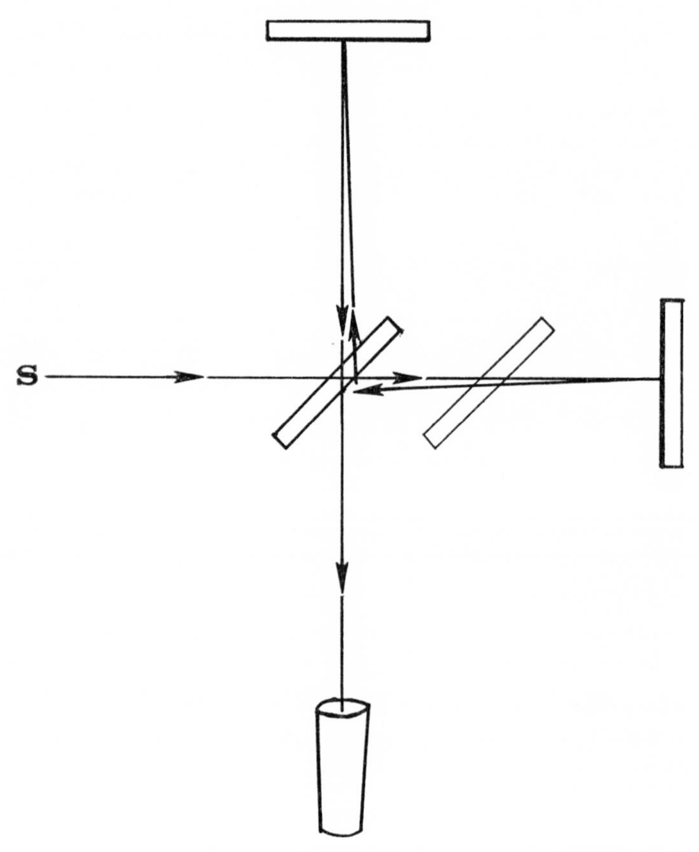

FIG. 2.1. The Michelson interferometer. *S* is the light source.

lations of the effects expected and the results obtained.

The interferometer is shown in Figure 2.1. The compensating plate is placed into one beam to compensate for the fact that the other beam makes an extra pass through the glass of the half-silvered mirror; thus the path lengths in glass for both beams are the same. When the observer looks with his eye alone, or with a small telescope as shown, he sees a pattern of dark and bright fringes which may be concentric circles (Fig. 2.2) or

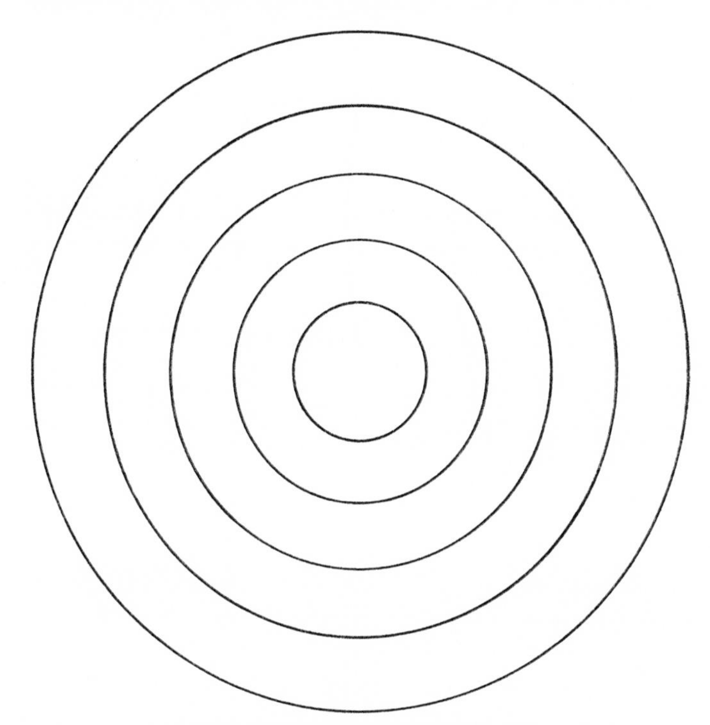

FIG. 2.2. The circular, centered fringe pattern seen upon looking through the telescope, if the totally reflecting mirrors are perpendicular to one another.

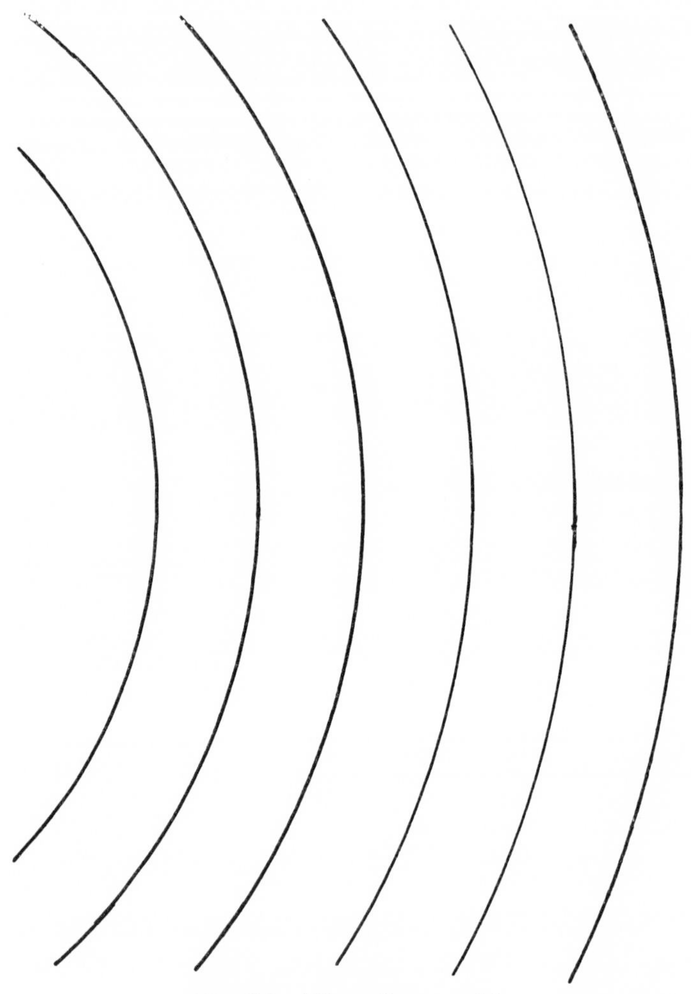

FIG. 2.3. The off-center fringe pattern seen when the reflecting mirrors are not precisely perpendicular to one another.

sections of circles of large radius whose center is far to the side (Fig. 2.3). The fringes are produced by the interference of the two beams of light which have traversed the two arms of the instrument. If either of the mirrors is moved, the fringe pattern changes; the pattern moves by one fringe for a path length change in either arm of one wavelength of the light. When the interferometer is used for a measurement of the Michelson-Morley type, the mirrors are not moved but the assumed flow of the ether through the instrument is expected to cause fringe shifts.

Probably the best way to understand the Michelson-Morley experiment is by consideration of an analogy. Imagine a river, flowing with speed v, and two boats, each of which is to travel a distance L from a starting point on the bank and then to return to that point. One boat goes first downstream a distance L, and then returns upstream. If the speed of the boat in the water is c, the time required to go down is $L/(c+v)$ and the time required to return is $L/(c-v)$. The total time is therefore

$$t = \frac{L}{c+v} + \frac{L}{c-v} = \frac{2Lc}{c^2 - v^2}.$$

The other boat goes directly across the river a distance L and then turns and returns to the starting point. It must point its nose slightly upstream so that it will not be carried down by the current; thus the component of its velocity which is directed across the stream is $\sqrt{c^2 - v^2}$ (Fig. 2.4). Its time for the round trip is thus

$$t = \frac{2L}{\sqrt{c^2 - v^2}}.$$

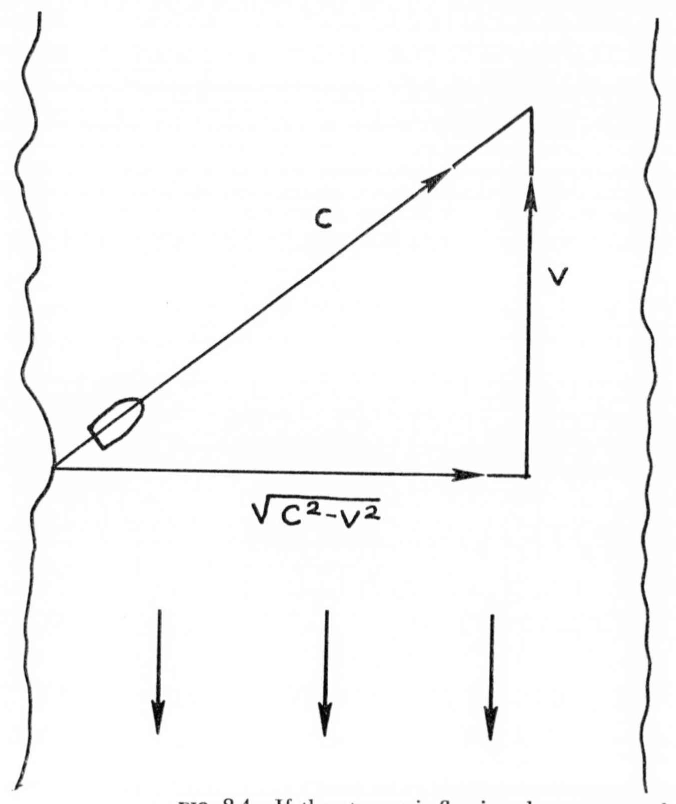

FIG. 2.4 If the stream is flowing down at speed v and the boat's speed in the water is c, the boat must point its nose upstream so that it has an upstream component of velocity equal to v. The cross-stream speed which results is only $\sqrt{c^2 - v^2}$.

The operation of the interferometer in the moving ether stream (moving as seen from the earth—actually, the earth and the instrument are moving through the stationary ether) is exactly analogous to the boats in the river; the light is the boats, and the ether is the water. Let us call the time for round-trip travel of the light parallel to the direction of ether flow $t\|$, and the time for travel perpendicular to that direction $t\perp$. Then these are given by the formulas derived above where L is the length of the arms (assumed equal) of the interferometer:

$$t\| = \frac{L}{c+v} + \frac{L}{c-v} = \frac{2cL}{c^2 - v^2}$$

$$t\perp = \frac{2L}{\sqrt{c^2 - v^2}}.$$

It is convenient to introduce the symbol β to represent the ratio v/c, and now if we divide numerator and denominator of these fractions by c^2 and c respectively and replace $(v/c)^2$ by β^2 the equations become

$$t\| = \frac{\frac{2L}{c}}{1 - \beta^2} = \frac{2L}{c}(1 - \beta^2)^{-1}$$

and

$$t\perp = \frac{\frac{2L}{c}}{\sqrt{1 - \beta^2}} = \frac{2L}{c}(1 - \beta^2)^{-\frac{1}{2}}.$$

If we expand each of the expressions in parentheses by the binomial theorem we get

$$t\| = \frac{2L}{c}(1 + \beta^2 + \cdots)$$

and

$$t\perp = \frac{2L}{c}(1 + \tfrac{1}{2}\beta^2 + \cdots).$$

The difference between the two times as a result of the ether drift is

$$\Delta t = t\| - t\perp = \frac{2L}{c}(\tfrac{1}{2}\beta^2) = \frac{L}{c}\beta^2.$$

The effect on the fringe pattern of the time difference is the same as if one arm had changed in length by $c\,\Delta t$; call this effective length change Δl, then

$$\Delta l = c\Delta t = L\beta^2.$$

The number of fringe shifts resulting from this effect is the ratio of the effective length to the wavelength of the light, or

$$n = \frac{\Delta l}{\lambda} = \frac{L\beta^2}{\lambda}.$$

We may now estimate the shift to be expected in the fringe pattern if the earth moves through the ether with the same velocity it has around the sun. The earth's orbital velocity is approximately 18 mi/sec, and the speed of light is approximately 186,000 mi/sec. Thus

$$\beta_{\text{orbital}} \equiv \frac{v_{\text{orbital}}}{c} \approx \frac{18}{186{,}000} \approx 10^{-4}.$$

In the interferometer used by Michelson and Morley, L was 1,100 cm. If we assume that the effective wavelength of the light used was 5,000 Å ($= 5 \times 10^{-5}$ cm), we may calculate the expected fringe shift to be

$$N_{\text{expected}} = \frac{L\beta^2_{\text{orbital}}}{\lambda} \approx \frac{10^3 \times 10^{-8}}{5 \times 10^{-5}} = \tfrac{1}{5} = 0.2 \text{ fringe.}$$

As the interferometer was rotated through 90°, the arms interchanged roles (parallel and perpendicular to the direction of ether flow) with the result that the expected shift is twice this much, or 0.4 fringe.

Instead of 0.4 fringe, the shift observed was no greater than 0.01 fringe. Let us see what limit that observation places on the speed of the earth through the ether. If we set $n = 0.01$ and use L and λ the same as before, we can calculate an upper limit on β:

$$\beta^2 \leqq \frac{nL}{\lambda} \approx \frac{10^{-2} \times 5 \times 10^{-5}}{10^3} = 5 \times 10^{-10}$$

from which

$$\beta \leqq 2.25 \times 10^{-5} = 0.225 \times 10^{-4} \approx \tfrac{1}{5}\beta_{\text{orbital}}.$$

A more exact calculation yields the value $\tfrac{1}{6}\beta_{\text{orbital}}$ as the limit set by the experiment. In other words, the Michelson-Morley experiment proved that the measurable velocity of the earth through the ether was less than $\tfrac{1}{6}$ its velocity about the sun. The result was unexpected and had profound consequences, but one might wish that the limit of motion through the ether could be set somewhat lower. The difficulty with this experiment, of course, is that the effect observed is proportional to $(v/c)^2$, and this is a very small number. The next experiment to be described makes use of an effect which is proportional to (v/c), and it yields a much lower limit on the ether-drift speed.

A method of measuring the speed of the earth through space to a precision much higher than that possible by the Michelson-Morley experiment was reported by Cedarholm, Bland, Havens, and Townes[4] in 1958; the final result of repeated experiments of this

---

4. J. P. Cedarholm, G. F. Bland, B. L. Havens, and C. H. Townes, *Phys. Rev. Letters* 1 (1958):342.

type was reported by Cedarholm and Townes[5] in 1959. The argument is slightly more difficult than that of the ether drift in the Michelson interferometer.

The instrument used is the ammonia maser. The ammonia molecule (NH_3) can exist in two states having slightly different energies; and when it goes from the higher state to the lower, it usually emits radiation to carry away the energy difference. This radiation for the two states used in the maser is at a frequency of approximately 24,000 megahertz. The molecule has the curious property, when it is in the higher energy state, of being repelled by a strong electrostatic field; whereas the molecules in the lower state are attracted by such a field. Thus if a beam of ammonia molecules is permitted to pass through a region with strong electrostatic field gradients, the molecules in the upper energy state tend to separate themselves from those in the lower energy state. In the arrangement used in the maser, the molecules pass down a channel formed by rods lying parallel to it, and the rods are charged to high potentials. The molecules in the higher energy state move toward the axis and leave the separator as a narrow beam, whereas those molecules in the lower energy state move away from the axis and hence are "defocused." In this way a beam of molecules is formed, almost all of which are in the higher energy state.

If nothing disturbs one of these molecules, it can exist in the upper energy state for a long time without spontaneously dropping to the lower state; but if radiation from one molecule which has made the transition strikes another molecule in the higher state, that molecule is likely to be stimulated to make the transition im-

---

5. J. P. Cedarholm and C. H. Townes, *Nature* 184 (1959):1350.

mediately and to emit radiation which is in phase with the radiation which struck it. In the maser the probability of such an interaction is increased by passing the molecules which have emerged from the selector into a resonant cavity, so that any radiation emitted by a molecule bounces back and forth within the cavity for a considerable period of time. During this time other molecules are constantly entering the cavity, and some are likely to interact with the radiation already present and to add to it. If the resonant frequency of the cavity is adjusted properly and the beam of incoming molecules is sufficiently intense, the device becomes an oscillator capable of producing enough radiation to maintain its own operation continuously and also to provide some radiation which can be drawn off to control a clock or other external device. The frequency of radiation emitted by the ammonia molecule is so precise that two masers can maintain their frequencies relative to one another for at least a year with an accuracy of 1 part in 10 billion. Stated another way, a clock controlled by an ammonia maser should gain or lose no more than one second in a few hundred years.

For the Cedarholm-Townes experiment two masers are used side by side with the ammonia molecules in them traveling in opposite directions. The entire apparatus is mounted so that it can be rotated 180°. The oscillations of the two masers are beat together so that even though the frequency is very high—23,870 megahertz—differences of frequency of much less than 1 hertz can be detected, and the effect sought should be seen in this difference.

Let us call the velocity of the molecules relative to the apparatus u and the velocity of the apparatus rela-

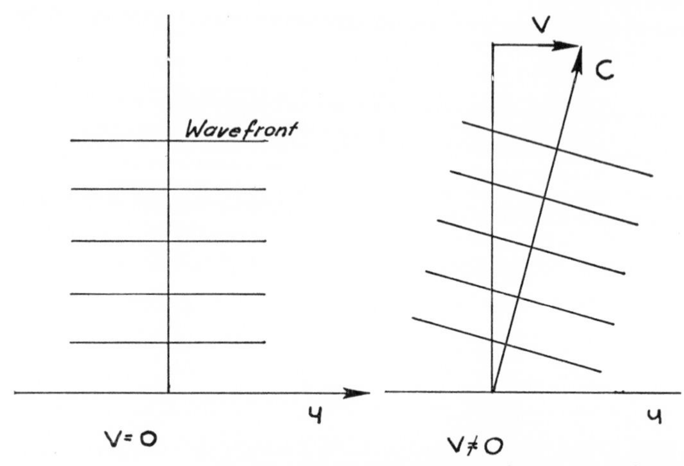

FIG. 2.5. If $v = 0$, the wavefronts are almost parallel to the direction of motion of the molecules; if $v \neq 0$, the wavefronts are tilted.

tive to absolute space v. Standing waves are produced in the apparatus by radiation emitted by the ammonia beam perpendicular to its direction of motion; and if $v = 0$ (no motion through space), that radiation perpendicular to the motion of the molecules has no Doppler shift. If, on the other hand, $v \neq 0$ and it is directed parallel to u, the waves which are to be parallel to the walls of the enclosure must be emitted slightly forward from the perpendicular; the angle made with the perpendicular is $\sin^{-1}(v/c)$. See Figure 2.5. Now the velocity u of the molecules does enter into the discussion, and there is a Doppler shift of the frequency because of it. The component of u seen by an observer at the wall of the enclosure looking in the direction from which the radiation reaching him was emitted is

$u(v/c)$, and the fractional Doppler shift in frequency produced by this velocity is

$$\frac{\Delta\nu}{\nu} = \left(\frac{u}{c}\right)\left(\frac{v}{c}\right) = \frac{uv}{c^2}.$$

If we put into this equation the speed of the ammonia molecules moving with thermal velocity, 0.6 km/sec, and for v the orbital speed of the earth around the sun, 30 km/sec, we find

$$\frac{\Delta\nu}{\nu} \approx \frac{0.6 \times 30}{9 \times 10^{10}} = 2 \times 10^{-10}.$$

The other beam of the two masers should have the same effect but in the other direction; and when the directions are interchanged by rotating the instrument through 180°, the effect should be doubled. One should expect to see a frequency change upon rotation, therefore, of four times this amount, 8×10^{-10}. Since the frequency used is approximately 2.4×10^{10} hertz, the expected frequency change is approximately 20 hertz.

The frequency shift actually found was less than 1/50 hertz. (This does not mean that a definite shift was found whose value was less than 1/50 hertz, but rather that experimental uncertainties were such that one can only say that if there was a shift at all it was less than this limit.) Since the frequency shift is proportional to (v/c), one can conclude that the experiment sets an absolute limit on the speed of the earth relative to absolute space of 1/1,000 its orbital speed around the sun. (Contrast this with the limit of 1/6 the orbital speed set by the Michelson-Morley experiment.) To eliminate the possibility that at a particular time the velocity of the earth through absolue space might have happened to be zero because of a combination of its orbital motion around the sun and the sun's motion, the experiment was repeated at three-month intervals throughout a year, during which time the earth had changed its velocity relative to the initial measurement by 60 km/sec (as it passed around the far side of the sun); the results were consistently the same. A useful way of stating the result is to say that at no part of the year was a velocity relative to absolute space observed as great as 3,000 cm/sec.

The third experiment to be considered sets even more stringent limits on the velocity of the earth relative to space. This is the Turner-Hill experiment[6], published in 1964, and it makes use of still another physical effect, the Mössbauer effect.

After a nucleus has undergone radioactive decay, the daughter nucleus is often left in an "excited state," which means that it still has some extra energy it must get rid of. This energy is usually given off in the form of a gamma ray, and for a particular nuclear species, the wavelength (or frequency or energy—all equivalent ways of describing the gamma ray) is constant and characteristic of the species. In addition to energy, however, a gamma ray carries momentum, and usually the nucleus must recoil when the gamma ray is emitted so that momentum can be conserved. Since the gamma is thus emitted from a recoiling nucleus, it shows a Doppler shift; and this is superimposed on the Doppler shift which arises from the fact that the nucleus is moving, either because it is in an atom in a gas or because it is in

6. K. C. Turner and H. A. Hill, *Phys. Rev.* 134B (1964):252.

a solid, where the atoms are all vibrating. The two effects on the gamma ray are (1) to shift its frequency from the original value and (2) to give the gamma rays which originate from a sample of material slightly different frequencies—in usual terminology to "broaden the line." This second effect can be greatly reduced by cooling the sample of material, so that the random motions of the atoms in the solid are reduced; but until Mössbauer discovered the effect which bears his name, there seemed to be no way of eliminating the first effect, the one which arises from the recoil of the nucleus.

Mössbauer discovered, however, that in certain cases, when a sample of material has been cooled to a very low temperature, part of the gamma rays which are emitted do not show the Doppler effect from the nuclear recoil. In these cases the momentum equal to that of the gamma ray is taken up by the entire sample of material instead of by just the one nucleus; and since the sample is very massive compared to the nucleus, the resulting velocity and hence Doppler shift is entirely negligible. Thus gamma rays are emitted which have a precisely defined frequency (or wavelength or energy), and just that frequency which corresponds to the energy change in the nucleus. The significance of the emission of a gamma with all the original energy becomes apparent when we consider the absorption of that gamma ray by another piece of material. If the gamma falls on a nucleus of the same sort it left and if it has just exactly the original energy, it is much more likely to be absorbed than under any other conditions. This is called "resonance absorption," and it occurs only when the emitter and absorber are "in tune"; that is, when the absorber is offered just the amount of energy to absorb which it could emit if the roles were interchanged. The usual materials for using the Mössbauer effect are Co^{57} as the emitter and Fe^{57} as the absorber. The Co^{57} nucleus decays to Fe^{57}, and the iron nucleus is left in an excited state. It gives up its energy in a gamma ray; and if that gamma ray falls on the nucleus of a similar Fe^{57} atom, that nucleus can absorb it. If the energy of the ray has changed slightly, the probability of its being absorbed is less. Thus if a beam of gamma rays from Co^{57} passes through a piece of Fe^{57}, some rays are absorbed or scattered under all conditions; but the absorption increases greatly when conditions are satisfied for resonance absorption. This increase in absorption is so sensitive to slight frequency changes that the Doppler shift arising from a motion of absorber relative to emitter of a fraction of 1 mm/sec can be detected. It is this sensitivity to relative motion which makes the Turner-Hill experiment possible.

Before describing the actual experiment, perhaps we should consider the Doppler effect briefly. As usually presented in elementary physics based on sound, the Doppler effect is likely to be described by an equation of the form

$$\nu' = \nu \frac{1 \pm \beta_o}{1 \mp \beta_s}$$

to cover the cases of approach and recession and of observer moving or source moving. When we deal with light, relativity theory assures us that the result should be the same whether the observer or the source is moving; for all that is physically significant is the relative motion, and the formula usually given is

$$\nu' = \nu\,(1 \pm \beta)$$

where the + sign goes with approach and the — with recession. This is the result obtained by assuming that Galilean relativity applies to the problem, and it is also the approximate result given by the theory of relativity correct to the first power in (v/c). The complete, correct formula, from special relativity, is

$$\nu' = \nu \frac{1 - \beta \cos \theta}{\sqrt{1 - \beta^2}}$$

where θ is the angle between the direction of v and the direction of propagation of the light. If we expand this by the binomial theorem we get

$$\begin{aligned} \nu' &= \nu (1 - \beta \cos \theta)(1 - \beta^2)^{-\frac{1}{2}} \\ &= \nu (1 - \beta \cos \theta)(1 + \tfrac{1}{2}\beta^2 + \cdots) \\ &= \nu (1 - \beta \cos \theta + \tfrac{1}{2}\beta^2 - \tfrac{1}{2}\beta^3 \cos \theta + \cdots). \end{aligned}$$

It will be seen that for light propagation along the line of motion ($\theta = 0$) and for v very small compared to c (β small), so that terms in β^2 and higher powers of β can be neglected in comparison with the first order term, this result agrees with the one derived from the Galilean transformation, that is, from classical theory. When these conditions do not apply, however, the differences between the predictions of the two theories are significant. If β is not small, then more terms of the expansion than just the first must be used. And even if $\theta = \pi/2$ so that $\cos \theta = 0$, there is a Doppler effect; this is called "the transverse Doppler effect." If the emitter is moving at right angles to the line of sight, the frequency observed is given by (neglecting fourth order terms)

$$\nu' = \nu (1 + \tfrac{1}{2}\beta^2).$$

This term appears because the clocks used by the emitter and the observer run at different rates as seen by the observer when the emitter is moving relative to him, regardless of the direction of the motion. This is a purely relativistic effect, unexplained by classical theory.

Now with this background in the Mössbauer effect and the Doppler effect, we can describe the experiment. A Co^{57} source was placed near the rim of a centrifuge wheel approximately 25 cm in diameter, and an Fe^{57} absorber was placed near the axis of the wheel. The source was below the absorber so that gamma rays passing from source through absorber could continue to a detector (a scintillating crystal) on the axis of the wheel. Light from the scintillator was carried through a lucite rod mounted on the wheel to a photomultiplier above the wheel. The pulses from the photomultiplier, counting the gamma ray photons which passed through the absorber, were divided into four groups; the total of each group was recorded by the electronic equipment attached to the photomultiplier. The division into four groups was made on the basis of the orientation of the wheel. As the wheel rotated, mirrors on its edge were used to operate electronic switches, so that during one quarter-revolution the counts from the photomultiplier went into one group, during the next quarter-revolution they went into a second group, and so on for the four groups.

As the wheel revolved, its speed alternately added to and subtracted from the speed of the earth relative to an absolute space; and this changing absolute speed should have caused a Doppler shift in addition to the one arising from the time shift from circular motion. The result should have been that the number of counts in the four groups would not be the same, and the variation in number should be correlated with the direction of motion of the emitter at the time counts for the group were taken. Furthermore as the earth rotated, the group which should have been expected to have an excess of counts changed because of the change in orientation of the laboratory with respect to the earth's orbital velocity.

No such effects were found, and this experiment gives an upper limit on the "ether-drift velocity" of 840 cm/sec, less than 1/1,000 the earth's orbital velocity.

The reader is likely to wonder, with some reason, why any physicist was still trying to measure the speed of the ether drift as late as 1961, when the last of these experiments was carried out. The answer is that no one now seriously considers the ether drift as a real effect, and absolute space is conceived in terms somewhat different from those that Newton probably had in mind. There are two reasons to continue to make this sort of measurement, however. One is that the impossibility of detecting absolute motion is one of the foundations of the theory of relativity; and if the foundation should be found false, the theory would have to be rethought. At the time Einstein proposed the theory, the best experimental evidence for the conclusion that absolute motion cannot be detected was provided by the Michelson-Morley experiment; and as we have seen, it set a limit which was not as low as we might wish. Since this presumed fact, that absolute motion is undetectable by any means whatever, is so important in the development of the theory of relativity, further work to test the assumption as carefully as possible is justified

There is a second reason, however, to reconsider

measurements of this sort. If, as we have suggested, inertial effects may arise from an interaction with distant matter, then velocity relative to that distant matter might produce detectable effects; therefore, we can suggest a reason for this sort of experiment without even mentioning absolute space or the ether. We may ask simply whether motion relative to the "fixed stars" produces effects which are locally significant. If it makes sense to speak of the center of mass of the universe, is there any interaction between the total mass of the universe and our test equipment which can give us information about our motion relative to that center of mass? From the experiments we have described the answer is that no such motion has been detected; hence a rather small upper limit can be set on the strength of any interaction which may exist. These experiments indicate that velocity relative to the center of mass of the universe cannot be detected; when we discuss Mach's principle we will consider the possibility of detecting acceleration relative to the center of mass. But before taking up Mach's principle and the origin of inertia, we must spend some time on a question which was deferred earlier—the relation of relativity to the questions regarding the reference systems in which the laws of mechanics are valid.

3
Reference Systems in Relativity Theory

NOW THAT WE HAVE SEEN the evidence indicating that absolute motion and absolute space cannot be detected experimentally, we must turn to the theory of relativity and consider what it has to say about the problem of choosing reference systems.

The term "relativity" suggests a modern theory, for one usually thinks of Einstein's theories which date from 1905. This is a somewhat oversimplified view of the idea of relativity, however, for Galileo recognized one form of relativity, as did Newton. Because Galileo first discussed this particular form of relativity in rather complete form, it is called "Galilean relativity."

Galilean relativity is the principle that if one reference system is suitable for the description of mechanical phenomena, any other system moving in a straight line with constant speed relative to the first system is an equally suitable system. Stated another way, no mechanical experiment can detect straight-line motion at constant speed relative to "absolute space." It was recognized that if one is inside the cabin of a ship, he cannot tell by dropping objects or by performing any experiment with masses and forces whether the ship is moving. From a mathematical point of view this result

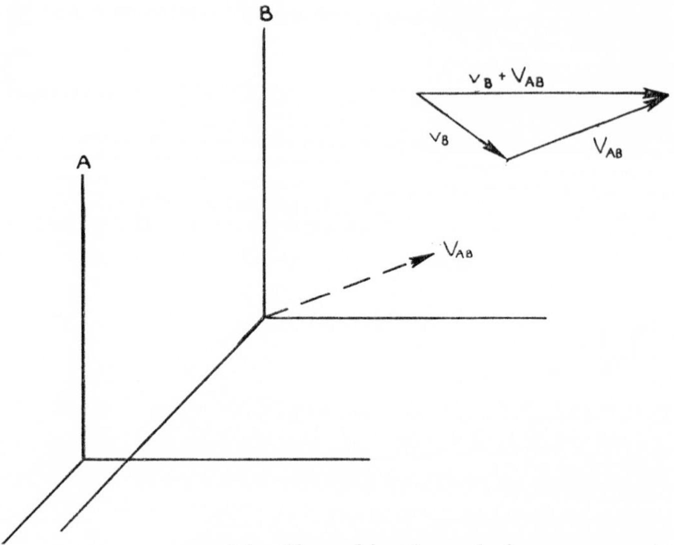

FIG. 3.1. If an object has velocity v_B as seen in system B, but system B has velocity V_{AB} as seen in system A, then its velocity as seen in system A is $v_B + V_{AB}$.

follows from the fact that the equations of motion (Newton's second law in equation form) contain the acceleration but not the velocity of the object under consideration. If reference system B has velocity V_{AB} relative to system A and if the velocity is a constant, then an object moving in system B with velocity v_B has velocity $v_B + V_{AB}$ in system A (Fig. 3.1), and the acceleration observed in the two systems is the same. Since only the acceleration enters the fundamental equations, one system is as good as the other for considering the motion of the object.

The need for absolute space in Newtonian physics is a logical need rather than a practical need. Physicists from Galileo to Einstein recognized that from a practical point of view any two reference systems moving with constant velocity relative to one another are entirely equivalent for mechanics experiments; the necessity for absolute space enters only in that some way is needed to find one reference system in which Newton's laws are correct—a system to which others can be compared. Mechanics could offer no method for finding the absolute reference system, however, because of the validity of the principle of Galilean relativity.

When the principles of electrodynamics were developed by Maxwell, the equations embodying them contained terms involving velocity instead of acceleration (for example, forces which are proportional to $v \times B$); hence these equations are not invariant under changes from one reference system to another moving with constant velocity relative to it. This is a fancy way of saying that if a phenomenon is described in terms of the velocities and accelerations measured in one reference system and then described in terms of the velocity and acceleration in another system moving with respect to the first, the equations used in the two descriptions look different. Usually one will include terms the other does not have, and those terms will involve the relative velocity of the reference systems. A simple example: When one views an electric charge which is at rest relative to him, he sees an electric field but no magnetic field; but if the charge is moving, it appears as a current and hence produces a magnetic field. Thus two observers, moving relative to one another, may look at a single charge and see entirely different effects—one seeing only an electric field, the other seeing a magnetic field also. Clearly the two observers are not equivalent. Thus it became meaningful to try to devise experiments to detect absolute motion by the use of electromagnetic phenomena—the propagation of light, for example.

The principle of special relativity which Einstein introduced into physics in 1905 extended the idea of relativity which had long been recognized in mechanics to include all physical phenomena. He interpreted the negative result of the Michelson-Morley experiment as an indication that two reference systems which are equivalent for mechanics are also equivalent for electromagnetism, and more generally, that if one reference system can be found in which the laws of mechanics are valid, then all laws of physics will be the same in that system as in another system moving with respect to it at constant velocity.

This principle of special relativity, if it is supported by experimental evidence (and it is!), removes forever the possibility of detecting absolute motion experimentally. Absolute space is beyond our detection by any sort of experiment, whether it involves mechanical, electromagnetic, or other phenomena. Note, however, this does not necessarily mean that motion relative to the matter in the distant parts of the universe cannot be detected, and it is possible to imagine mechanisms of interaction between masses which would produce detectable effects as a result of motion. It is this sort of effect which the Turner-Hill experiment attempted to find.

After we admit that the experimental detection of absolute space is impossible, we may still ask about its logical status: Must we still postulate the existence of an absolute reference system in order that relativity theory, or more specifically relativistic mechanics, may be a logically complete system? The answer is a qualified No. Let us consider in some detail the logical foundations of the mechanics of special relativity.

The question we originally asked has been changed, and it is important to realize in what way. We began by asking about the reference system in which Newtonian mechanics is valid, and now we are asking about the system in which relativistic mechanics is valid. These are two theories, and we have posed two questions which may have different answers.

The important differences between Newtonian and relativistic mechanics actually have no bearing on the choice of reference systems. One difference is that Newtonian mechanics assumes that signals can travel with infinite velocity, whereas relativity theory assumes that no signal can travel with a velocity greater than that of light. Thus using Newtonian theory, one writes that the force of attraction between two masses is $F = G\,(m_1m_2/r^2)$ and is directed along the line joining them. This implies that no matter how far apart the masses may be nor how rapidly they may be moving, the force acts along the line which joins them now. Relativity theory assumes that time is required for the effect of one mass to be transmitted to another; consequently, the attractive force at any instant is directed along the line toward the point occupied by the other mass a time r/c ago, when the influence now being felt was emitted. In this, as in many respects, Newtonian mechanics is the limiting case of relativistic mechanics, corresponding to an infinite light-travel velocity. Other important differences between the two theories are the identification of inertial properties with energy and the related identification of mass with energy through the famous equation $E = mc^2$ in relativistic theory, for which no corresponding identification exists in Newtonian theory, and the very important fact that relativistic theory (special theory of relativity) does not attempt to deal with gravitational fields. In fact, special relativity is strictly applicable only in the absence of gravitational fields.

The first answer to the question about absolute reference systems in relativity is that no special reference system is logically required because the theory is intentionally devised to make such a requirement unnecessary. All Galilean reference systems are treated on the same basis. But the reader will notice that we have slipped in a familiar word there—Galilean. What the theory of relativity really does is tell us that if we can find one Galilean system—that is, one reference frame in which Newton's first law is valid—it and all reference systems moving with constant velocity relative to it are equally good for formulating all of physics. How does this differ from Newtonian mechanics and Galilean relativity? It differs in that the old relativity applied only to mechanics; Einstein's theory extends the relativity principle to all physical principles and laws. In the course of working out the consequences of the principle of relativity as applied to the propagation of light, Einstein was led to propose profound modifications of our concepts of time and distance; but still, in this theory, he tied it to Galilean reference systems. It is somewhat like a new recipe for cooking chicken—the method of handling the meat differs from the old recipe, but it still begins "First catch a chicken." For strict application of the special theory of relativity we must find the same sort of reference system which Newtonian mechanics required; if we cannot find an ideal system, we are forced to make the same sort of approximations we made in Newtonian theory.

Since the special theory of relativity is strictly valid only in the absence of gravitational fields, no ideal reference system is known or is likely to be found, and one always has to work with approximations. Usually, however, this fact is completely unimportant in applications of the theory, just as the fact that the surface of the earth is not a Galilean reference system and hence not an appropriate place to use Newton's laws of motion is of no importance to the man computing the acceleration of an automobile acted on by a known force. It is difficult to see much difference logically between the requirements of Newtonian theory and the special theory of relativity in their demands on reference systems. Both are used in less than ideal situations, but the ideal system for one is also the ideal system for the other. The formulation of the special theory of relativity is such as to discourage the search for experimental evidence of absolute motion, but at the same time the theory is devised on the assumption that ideal Galilean reference systems can be found or at least that their discussion makes sense.

The special theory of relativity is strictly applicable only in a region of space where there are no gravitational fields, and the theory cannot deal with gravitation. It was to eliminate this restriction that Einstein devised the general theory of relativity.

To consider the implications of the general theory of relativity for the problem we are discussing, it is entirely unnecessary to be familiar with the mathematics of the theory; but we must have some understanding of the underlying reasoning. The fundamental idea is that the effects of gravitation can be described by motion in a curved space without the effects of a force. An analogy may help: If two ants start from points 3 inches apart on the equator of a globe and each travels straight north, they will find after crawling some distance that they are no longer 3 inches apart. One possible explanation is that some mysterious force is pulling them together; this is the explanation of gravitational effects given by Newton. Another possible explanation is that no force acts, but that they are on a curved surface. This, of course, is the explanation we would give for the ants on the globe, and an extension of this idea to four dimensions is the explanation Einstein gave for gravitation. If we should hold two bricks 10 feet apart and 100 feet above the ground and then drop them, careful measurement would reveal that they were less than 10 feet apart when they struck the earth. Newton said they are both attracted to the center of the earth; Einstein said that the mass of the earth causes the space-time in its vicinity to be curved so that if the two bricks each move on a "straight" line (actually a geodesic) in space-time, they will approach one another. Since this explanation is to explain the same observations as Newton's theory which postulated a force between bodies proportional to the product of their masses, the mass of the bodies must enter into the theory in some way, and it does. The mass of the matter and the effective mass of the radiant energy in a region determines the curvature of space-time in that region. Thus near the sun the curvature is more severe than near the earth; and in a region where there is very little matter or radiation, the curvature is almost zero. In that case we say that the space is "flat"; it is Euclidean like the space we customarily use in our laboratories.

The possibility of identifying gravitational fields with curvature of space-time rests upon the fact that the proportionality of inertial to passive gravitational mass is the same for all materials; for if gravitation is to be considered a property of space, then all matter must respond to it in the same way. One can identify three kinds of mass: inertial, passive gravitational, and active gravitational. Inertial mass is the measure of resistance to acceleration; it is the quantity represented by m in the equation $F = ma$. Active gravitational mass is the property of matter which causes it to attract other matter, and passive gravitational mass is the property which causes matter to be attracted. One can say that matter produces a gravitational field, in Newtonian terms, and the field is given by an expression of the form Gm/r^2. This m is the active gravitational mass. In the presence of that field a piece of matter experiences a force of magnitude Field $\times$ m, and this m is the passive gravitational mass. By Newton's third law the active and passive gravitational masses must be equal, and hereafter we shall refer to either of them as merely the gravitational mass.

The relation between inertial and gravitational mass must be determined by experiment, and Newton seems to have been the first person to carry out experiments for this purpose. He used a pendulum to determine whether the period is independent of the material used to construct the pendulum. If one derives the equation for the motion of a simple pendulum, beginning from the fundamental equation

$$\begin{aligned}&(\text{inertial mass } m_i) \times (\text{acceleration})\\ &= (\text{grav. mass } m_g) \times (\text{grav. field})\end{aligned}$$

he arrives at a relation

$$T = 2\pi \sqrt{\frac{m_i}{m_g}} \sqrt{\frac{l}{g}}.$$

The fact that the period is independent of the nature of the material used for the mass on the pendulum implies that the ratio m_i/m_g is the same for all the materials used. This is the result found by Newton, but the method is not very sensitive.

A far more sensitive test was carried out in 1889 by Baron von Eötvös, and a refined form of this experiment was carried out at Princeton University by Roll, Krotkov, and Dicke[1] in the early 1960s. The principle of the experiment can be seen by an analogy. Suppose that within a satellite traveling around the earth there is a beam, supported by a fiber to make a torsion balance, and on the ends of the beam are masses of different materials. Each end of the balance is attracted by the earth, and the force of attraction is proportional to the gravitational mass of the end object. At the same time each end experiences an inertial reaction equal to its inertial mass times the centripetal acceleration. If these two "forces" (one a real force and the other an inertial reaction) are not the same for the two ends of the balance, the beam will twist. If the satellite is spinning so that the orientation of the beam relative to the earth is constantly changing in a periodic way, any

1. P. G. Roll, R. Krotkov, and R. H. Dicke, *Ann. Phys.* 26 (1964):442.

twist resulting from the lack of equality of forces on the ends will be periodic, going through a cycle for each turn of the satellite.

In the Princeton version of the experiment the beam is replaced by a triangular platform, and instead of two masses three were used—two cylinders of aluminum and one of gold. The platform (6 cm on a side) was suspended by a quartz fiber in a well in the ground and in a tank from which most of the air was pumped and in which the temperature was held as nearly constant as possible. Observations of the position of the platform were made entirely remotely, by means of an optical and electronic sensing and recording system. The sun replaced the earth and the earth replaced the satellite. The gravitational attraction which was important was that of the sun for the aluminum and the gold; and if the ratios of inertial to gravitational masses had not been the same for aluminum and gold, the platform would have tended to twist, the direction of twist depending upon the orientation of the platform relative to the direction of the sun. But the rotation of the earth on its axis changed that orientation by 360° every 24 hours, so if a twist of the platform had been produced, it would have had a 24-hour periodicity. The equipment was operated for months, and the conclusion of the experimenters was that any difference between the ratios of inertial mass to gravitational mass for aluminum and gold was less than 1 part in 10^{11}.

Aluminum and gold were chosen for better reasons than easy availability, and from the equality of their mass ratios a rather general conclusion can be drawn. The mass of an atom depends upon several effects—number of nucleons present, nuclear binding energy, number of electrons, electronic binding energy, and so forth. Gold and aluminum differ sufficiently in all these respects so that one can conclude that the part of the mass arising from each effect must obey the equality: inertial mass = gravitational mass. Thus it is reasonable to generalize that for any materials one might use, the same result would be found.

The null result of the Eötvös experiment does not prove that general relativity is correct or that gravitation can be described in terms of geometry alone, but the result is a necessary condition for either conclusion. If the experiment had not found equality of inertial and gravitational mass, general relativity would have been proved incorrect; and the sort of arguments used to arrive at the general theory would have been proved inappropriate.

There are two things about the theory of general relativity which we must emphasize: (1) It is formulated in such a way that any coordinate system is as good as any other system as a basis for describing the laws of nature, that is, no Galilean system is needed, and (2) at any point in a sufficiently small region a Galilean system can be imposed. This is equivalent to saying that in a sufficiently small region the space looks flat, just as on the earth, in a sufficiently small area, we can neglect the fact that the surface is curved and use Euclidean geometry to describe triangles and other geometric figures. There is also a third aspect of the theory which is important, and we shall describe it in a moment.

These two properties of the theory would seem to end, once and for all, any discussion about absolute space or preferred Galilean reference systems. It is agreed that no such systems can be found experimentally, and now general relativity formulates our description of nature in such a way that none is needed. Hooray! We are at the end of the road! But then one doubt sneaks back in, the third aspect of the theory mentioned above—if we go very far away from all masses, the curvature of space becomes zero; and we can describe all space at this great distance from masses in terms of a Galilean reference system. Near the masses space is curved, but at great distances it becomes uniform. Have we not here reintroduced absolute space? Some relativists think that we have, for example, Synge, who writes:

> In Einstein's theory, either there is a gravitational field or there is none, according as the Riemann tensor does not or does vanish. This is an absolute property; it has nothing to do with any observer's world-line. Space-time is either flat or curved. . . . The Principle of Equivalence performed the essential office of midwife at the birth of general relativity, but, as Einstein remarked, the infant would never have got beyond its long-clothes had it not been for Minkowski's concept [the geometrical way of looking at space-time]. I suggest that the midwife now be buried with appropriate honours and the facts of absolute space-time faced.[2]

Somewhat the same sort of idea was expressed by Einstein in an address delivered May 5, 1920, at the University of Leyden, titled "Ether and the Theory of Relativity," which ended thus:

> Recapitulating, we may say that according to the general theory of relativity space is endowed with physical qualities; in this sense, therefore, there exists an ether.

2. J. L. Synge, *Relativity: The general theory* (Amsterdam: North-Holland Publishing Co., 1960).

> According to the general theory of relativity space without ether is unthinkable; for in such space there not only would be no propagation of light, but also no possibility of existence for standards of space and time (measuring-rods and clocks), nor therefore any space-time intervals in the physical sense. But this ether may not be thought of as endowed with the quality characteristic of ponderable media, as consisting of parts which may be tracked through time. The idea of motion may not be applied to it.[3]

After this cheerful note what can be said in summary is the answer to our original question, What is the proper reference system for the application of Newton's laws? First, we can say emphatically that Newton's laws are only approximations, and the laws of special relativity are the correct ones to be applied in any small region of space. Second, we may assert equally emphatically that the possibility of finding any preferred reference system (absolute space) by experiment seems to be absolutely nil. Third, we know that in a small region of space it is always possible to find a coordinate system in which special relativity is valid, that is, a Galilean system. And fourth, we must conclude that the extent to which the concept of absolute space as a logical element of the theory has been exorcised is a matter of opinion; there is no certain answer yet. At the same time, although absolute space has been consciously eliminated from the main body of the theory, it may have crept in by the back door in the form of the boundary conditions which must be used at great distances from matter.

One aspect of motion with respect to distant mat-

3. A. Einstein, *Sidelights on relativity* (London: Methuen and Co., Ltd., 1922).

ter or to the radiation which fills all space perhaps should be mentioned in order to avoid misunderstanding. According to the theory of relativity no one can perform an experiment in a closed car of a train moving at constant velocity along an absolutely smooth track which will tell him whether the car is moving with respect to the earth (assuming that the earth is an inertial system). This does not preclude the possibility, however, that he will determine whether he is moving by opening a window and looking out. In the same way it may be possible to observe very distant matter and find that the Doppler shift of the light reaching us depends upon direction of sight and hence to deduce that we are moving relative to that matter. In the same way, it may be possible to determine from very careful observations of the microwave radiation in the space around us that we are moving relative to that radiation. (It may look different as we look in different directions.) This would mean that we are moving relative to the bulk of matter which last scattered the radiation in the distant past. This would tell us something about our motion relative to other matter in the universe, but it would not tell us that there is an absolute space.

4
Inertia and Mach's Principle

THE KEY IDEA in our consideration of the nature of inertia is Mach's principle. Although the idea is implicit in Mach's discussions of inertia, he did not formalize it by the appellation "principle," and in fact he never gave a precise statement of it. The name "Mach's principle" and its proposal as an important consideration in the foundations of mechanics were due to Einstein. There is no universal agreement as to what should be included in Mach's principle; but as we shall use it, it is the statement that the inertia of bodies depends upon some interaction with the other mass of the universe. If the principle is valid, it would be reasonable to expect that if the total mass of the universe were greatly reduced, the inertia of all objects would be also reduced, perhaps in a proportional way.

As indicated in Chapter 1, one of the most convincing arguments for Mach's principle is based on the presumption that only relative motion is physically real. To return to the example of the rotating bucket of water, the analysis normally applied in mechanics treats the rotation of the water as if it were absolute; and on the assumption that the water is moving, the mathematical analysis correctly predicts the curvature of the surface. If one insists, however, that only rotation relative to the remainder of the universe is relevant and the same curvature of the water should be predictable from the assumption that the water is not moving but that the remainder of the universe is rotating around it, he finds that neither classical nor relativistic mechanics is able to provide an analysis which is satisfactory. It was on this point that Mach attacked classical mechanics, arguing that the lack of symmetry between moving water–stationary universe and stationary water–moving universe represented a defect in the formulation of the science. This argument impressed Einstein greatly, and one of his goals in the development of general relativity was the incorporation of the principle into physical theory; he was not entirely successful, a fact that he himself recognized.

Mach's principle is not difficult to state in words, and plausible arguments for its validity can be formulated rather easily; but incorporating it into a mathematical theory has proved to be surprisingly difficult. One method which might be used was suggested in 1952 by a British cosmologist, D. W. Sciama.[1] The theory which Sciama proposed was incomplete and in fact quite unsatisfactory, but it was offered as an indication of the method which might be used to devise a more complete theory expressing Mach's principle. The demonstration theory was worked out using relatively simple mathematical tools; the more complete theory, if it could be developed, would require much more involved mathematics.

Sciama patterned his theory after Maxwell's theory

---

1. D. W. Sciama, *Monthly Notices Roy. Astron. Soc.* 113 (1953):34; *The unity of the universe* (New York: Doubleday, 1959).

of electromagnetism, defining a potential analogous to the electrostatic potential and introducing the effect of motion in a manner somewhat as magnetism accounts for the effect of moving electric charges. The result is an expression which relates the acceleration of a body which is attracted by a large mass to the total mass of the universe, the radius of the visible universe, and two other constants—the gravitational constant G which appears in Newton's formula and the velocity of light. The theory expresses Mach's principle in that it predicts that the inertia of a body depends upon the mass of the universe around it. It also predicts that very distant matter has vastly more effect than matter nearby (because the amount at great distance is much greater than the amount nearby) and hence explains why the effect of outside mass on the inertia of a body being studied is not readily apparent. Unfortunately, however, even though the theory developed by Sciama appeared promising, its elaboration in more accurate mathematical formalism has not yet been successful.

The most promising attempt to build Mach's principle into a theory appears to be a modification of Einstein's general theory of relativity proposed by Brans and Dicke; this will be discussed in the next chapter. We turn now to some experimental approaches to Mach's principle.

Several attempts to relate experimental results to Mach's principle have been inspired by a paper by G. Coconni and E. E. Salpeter[2] published in 1958. Their suggestion was that if the inertial mass of a body is determined by some interaction with all the other mass in the universe and that other mass has an aniso-

2. G. Cocconi and E. Salpeter, *Nuovo Cimento* 4 (1958):646.

tropic distribution, the inertial mass may be anisotropic; that is, it may be different if measured by means of accelerations in different directions. One presumes, for lack of information to the contrary, that the universe viewed on a sufficiently large scale is isotropic so that the mass at a great distance is uniformly distributed about us. But we know that we are not at the center of our galaxy, and consequently its mass is not uniformly distributed about us. If we only knew how the inertia of a body depends upon the external mass, we could estimate the importance of this galactic anisotropy.

Coconni and Salpeter proposed the simplest possible dependence, that the contribution Δm to the inertial mass m of a body which is produced by a mass ΔM at a distance r from the body is proportional to ΔM and inversely proportional to some power of r, so that

$$\Delta m \;\alpha\; \frac{\Delta M}{r^{\nu}}$$

where

$$0 < \nu < 1.$$

It is quite conceivable that the inertia of a body might depend upon the amount and distance of mass around it without having any dependence upon the direction to the mass; and if this were the case, the relationship just written would express everything one could say about the dependence of inertia on distant mass. This would not be a result easily subjected to test, however, so we consider the possibility that the inertial mass also depends upon the angle between the acceleration of the body and the direction of r to the distant mass. Thus we should expect the mass which appears in the equation $F = ma$ to be different if the body were accelerated toward the center of our galaxy than if it were accelerated at right angles to the line joining us to the center of the galaxy. And if the acceleration vector made an angle of 30° with the direction to the center, we should expect still another value for the mass. We might add another factor to the expression given earlier so that

$$\Delta m \ \alpha \ \frac{\Delta M}{r^{\nu}} \ f(\theta)$$

where $f(\theta)$ gives the dependence upon angle, and θ is the angle between the direction of acceleration and the line between Δm and ΔM. We are now expressing the relationship generally, not restricted to the mass of the galaxy.

For $f(\theta)$ we shall take the simplest function which satisfies some elementary symmetry conditions. The function must be an even one, that is $f(\theta) = f(-\theta)$, because it would be nonsense physically to suppose that the effect of one mass upon another depends upon which way one chooses to measure angles, so that a particular angle is positive or negative. The simplest function which satisfies this demand is the cosine. But there is another consideration–we shall suppose that the effective mass for acceleration toward the ΔM is the same as for acceleration away from it; in other words, we want the function to be the same for $\theta = 0$ as for $\theta = \pi$. Cosine θ does not satisfy this criterion, and the simplest function which does is $\cos^2 \theta$. Coconni and Salpeter did not actually use merely $\cos^2 \theta$, but a function which occurs very often in mathematics and physics called the second order Legendre polynomial, designated $P_2(\theta)$ and defined to be

$$P_2(\theta) = \frac{3\cos^2\theta - 1}{2}.$$

We may now write our assumption about the contribution made by a distant mass to the inertial mass of an object in the laboratory as

$$\Delta m \propto \frac{\Delta M}{r^\nu} P_2(\theta).$$

Now we shall apply this assumption about the contributions to the mass of an object to a simplified model of the universe. We imagine all the mass of the universe except for our galaxy to be uniformly distributed about us, and we further imagine all the mass of the galaxy to be concentrated at its center. We now suppose that the inertial mass of a test body we plan to examine can be divided into two parts, one arising from interaction with the mass of the universe outside the galaxy and the other arising from interaction with the mass of the galaxy. We will call that part of its mass arising from the effect of very distant mass m_0 and the part arising from the effect of the galaxy Δm, and the total mass of the body is $m = m_0 + \Delta m$.

Now we would like to determine something about the ratio $\Delta m/m_0$. A reasonable assumption about the mass M of the universe outside our galaxy is that it is uniformly distributed around the earth out to some radius R which is the effective radius of the universe. Then

$$m_0 \; \alpha \int_0^R \frac{4\,\pi r^2\,\rho\,dr}{r^\nu} = \frac{4\,\pi\rho R^{3-\nu}}{3-\nu}$$

where ρ is the density of mass in the universe. And if the mass M_0 of our galaxy is located a distance R_0 from us,

$$\Delta m \;\alpha\; \frac{M_0}{R_0{}^\nu}.$$

The ratio of these two is

$$\frac{\Delta m}{m_0} = \frac{M_0}{R_0{}^\nu}\,\frac{(3-\nu)}{4\,\pi\rho R^{3-\nu}}.$$

Now what can be said about the value of ν? That it must lie between 0 and 1 is apparent from these considerations: If it were less than 0, the influence of mass would increase with its distance. In other words, a given mass would have more influence on the inertia of a body on the earth if it were very far away than if it were near. That is unreasonable. On the other hand, if the value of the constant were greater than 1, relatively nearby objects such as the sun would have the predominant effect on inertia; and this is contrary to observation. Hence ν must be a positive number less than 1.

Reasonable values for the constants appearing in the equation are $M_0 = 3 \times 10^{44}$g, $R_0 = 2.5 \times 10^{22}$cm, $R = 3 \times 10^{27}$cm, and $\rho = 10^{-29}$ g/cm^3. With these values, we may compute the extreme values of the ratio by taking $\nu = 0$ and 1. If $\nu = 0$, $\Delta m/m_0 = 3 \times 10^{-10}$; if $\nu = 1$, $\Delta m/m_0 = 2 \times 10^{-5}$.

One must next find a physical system in which very

small differences in the mass might be detected as the system is rotated so that the mass moves along a line toward the center of the galaxy and then at right angles to that line. One might consider, for example, a mass between two stretched springs so that the mass can be made to oscillate along the line of the springs whether that line is vertical or not; in this way the oscillation could be directed along the line to the center of the galaxy, and then the system could be rotated so that oscillation took place at right angles to that line. The period of such oscillation is proportional to $\sqrt{m}$; so if the periods when the mass has the values m and $m + \Delta m$ are compared, we are comparing $\sqrt{m_0}$ to $\sqrt{m_0 + \Delta m}$, and the relevant ratio is

$$\frac{\sqrt{m_0 + \Delta m}}{\sqrt{m_0}} = \sqrt{1 + \frac{\Delta m}{m_0}} = (1 + 2 \times 10^{-5})^{\frac{1}{2}} \approx 1 + 10^{-5}$$

where the largest value for $\Delta m / m_0$ has been used. If the smallest value is used, the change in period would be of the order of 10^{-10}. The chance of seeing such small changes in periods of the order of one second is very slight. Macroscopic systems have enough damping and enough disturbances to make such sensitive measurements of period almost impossible.

Four types of systems have been used to make measurements in which mass anisotropy might be expected to be seen. In each case the actual procedure did not involve rotating equipment to point it toward the center of the galaxy and then away from that line; but rather the equipment was fixed on the earth, and the movement of the earth as it rotates on its axis was used to change the orientation of the equipment relative to the galactic center. One could expect a 12-hour periodicity if the mass anisotropy is to be detectable. Full explanation of the arguments in these four cases cannot be given within the limitations of this book, but qualitative explanations can be given.

The first of the four is a quartz crystal used in an oscillator. Quartz crystals are used as the control elements in oscillators which must keep very precise frequencies, because the crystal vibrates mechanically and this mechanical vibration determines the frequency of electrical oscillation in the attached circuit. The system works in this way: An electronic circuit generates an oscillating current, and this is made to produce an oscillating potential difference between metallic plates attached to the sides of a quartz crystal. As an electric field is applied across the crystal, it changes size; so that as the field fluctuates, the crystal swells and shrinks, and its changes in size in turn produce a potential difference across it. If the electric field is applied at the frequency at which the crystal is naturally resonant, its changes in size are relatively large and it in turn helps increase the electric field. If, on the other hand, the applied field is at a frequency other than the resonant frequency of the crystal, its response is small and it makes a small contribution to the field. Thus if the electrical circuit produces a signal near to the resonant frequency of the crystal, there is a strong tendency for the circuit oscillation to go to that resonant frequency and to remain there. Since the resonant frequency of the crystal depends upon its mass, one should expect a variation in the frequency of resonance as the crystal is turned so that the direction of movement of its parts is along the line toward the galactic axis or perpendicular to that line.

One could rotate such an oscillator and look for variations in frequency, but rotation in the laboratory introduces a number of problems which may mask, or at the least complicate, the effects being sought. For example, the earth's magnetic field may affect the frequency of the oscillator; and if the laboratory equipment is rotated, the horizontal component of the earth's field through it changes, producing unknown variations. Rotation may also introduce mechanical strains and effects of the immediate environment, for example, the electric and magnetic fields arising from the laboratory power facilities. Many of these effects can be eliminated by leaving the equipment fixed in the laboratory and watching for frequency changes as the rotation of the earth carries the equipment around during a 24-hour period.

Changes in frequency can be detected only by comparing the output of the oscillator whose frequency is being observed with a clock whose frequency does not change. But if the rotation of the earth is expected to produce a frequency change in one system, how is one to find a comparison clock which may not show the same effects? Crystal oscillators provide the frequency standard of many extremely precise clocks, and such a clock obviously could not be used for this measurement because one should expect that both crystals would be affected by the earth's rotation in the same way, so that they would show no frequency difference. There exists a solution, however, in the form of a cesium beam clock which makes use of transitions within the cesium atom between two states which are not expected to show any effect of mass anisotropy. Thus it seems reasonable to suppose that even if the mass of the cesium atoms depends upon the interaction with the mass of the galaxy and with the more distant mass of the universe, the frequency of the radiation associated with the energy changes within the atom would not be affected by the rotation of the apparatus because of the spherical symmetry of the atoms.

Measurements of the sort suggested were carried out by Essen, Parry, and Steele[3] in 1960, and they found no evidence of mass anisotropy. They calculated that their measurements set an upper limit on the effect of $\Delta m/m_0$ of approximately 1 part in 10^9.

A more sensitive test for mass anisotropy is provided by the atomic Zeeman effect. The principal observation which one makes of atoms is energy changes—the energy absorbed or released as the atom makes some internal change. The atom is said to change from one energy state or energy level to another; and from observation of the amounts of energy released or absorbed, it is possible to infer the energy (compared to some reference level) associated with various states of the atom. These states can also be related to the movements of the electrons outside the nucleus. When an atom is placed in a magnetic field, some energy levels are shifted slightly so that the amounts of energy required for given transitions are changed by small amounts. The most common effect is to split single energy levels into several closely spaced levels, so that a transition which previously involved a particular amount of energy is replaced

3. L. Essen, J. V. L. Parry, and J. McA. Steele, Proceedings of the Institute of Electrical Engineers 107B (1960):229.

by several transitions each with very slightly differing amounts of energy. This splitting is the Zeeman effect and is purely the effect of the magnetic field on the energy states of the atom.

When an external magnetic field defines a spatial direction, certain energy states of the atom are associated with the movement of an electron in the atom along the direction of the field; and other energy states are associated with movements which either are in the plane perpendicular to the field or are not correlated with the field at all. If the mass of the electron is different as it moves along different directions, then the energies of these states will be affected by the mass anisotropy in addition to the magnetic field; and the effect will depend upon whether the electron moves along a preferred axis or not. Thus if one studies a transition from a state for which the motion is along the direction of the magnetic field to a state having no preferred direction, he should find the energy difference between the states changing, as the direction of the field varies with respect to the line to the center of the galaxy.

To oversimplify the picture for the sake of conveying the essential idea—when a properly chosen atom is placed in a magnetic field, the outer electrons which are responsible for certain energy changes oscillate along the direction of the field. The energy which those electrons have can be measured with great precision because it is compared with the energy they have when they are moving in other directions that are not affected by the magnetic field. As the earth turns, the angle between the magnetic field and the line to the center of the galaxy changes; and if the effect of mass anisotropy is to be observed, the energy of the electrons oscillating along the field direction should change. The changing energy should appear as a changed frequency of the radiation emitted or absorbed by the atom, and again a 12-hour periodicity is expected.

Detailed calculation of the frequency changes to be expected are beyond the scope of this book, for they require consideration of the perturbations of energy states of atoms, and the methods of quantum mechanics must be used. The results of the measurements can be stated very simply, however. Experiments carried out by Beltran-Lopez and Robinson[4] and by Harvey, Kamper, and Lea[5] on the Zeeman effect in both chlorine and oxygen give upper limits for the effect of mass anisotropy of $\Delta m/m_0 < 10^{-10}$.

A third process which might show the effect of mass anisotropy is the Mössbauer effect. According to the shell model of the nucleus one of the states of Fe^{57} involved in the transition usually used to demonstrate the Mössbauer effect may exhibit mass anisotropy, with the direction along which the mass is maximum correlated with the magnetic field at the nucleus. Thus if the atom is placed in an external magnetic field which tends to align the nucleus with it, the effect of changing mass can be sought as the direction of the magnetic field varies relative to the line to the center of the galaxy. This is very similar to the argument in the case of the Zeeman effect except that there the effect is in the electrons in the outer part of the atom, and the Mössbauer effect occurs within the nucleus, with the nucleons moving instead of electrons. The effect of mass anisotropy in the Mössbauer effect was sought by a

4. V. Beltran-Lopez and H. G. Robinson, *Phys. Rev.* 123 (1961):161.
5. J. S. M. Harvey, R. A. Kamper, and K. R. Lea, *Proc. Phys. Soc. (London)* 76 (1960):979.

group of six men at the University of Illinois with negative results. They concluded that $\Delta m/m_0 < 5 \times 10^{-16}$, although they emphasized that this estimate is made assuming the validity of the shell model of the nucleus; and the model yields slightly incorrect predictions for another quantity which is related to the mass anisotropy.

The most sensitive measurement looking for mass anisotropy was made using the technique of nuclear magnetic resonance. Experimental work was done by Beltran-Lopez, Robinson, and Hughes[6] at Yale and independently by Drever[7] in Scotland. Again the argument depends upon the asymmetry of a nucleus, in this case the Li^7 nucleus, which in the ground state has one proton outside a spherically symmetric closed shell. In an external magnetic field the nucleus tends to align itself with the field in such a way that the proton moves chiefly along the field direction. If the kinetic energy of the proton is affected by an interaction with the mass of the galaxy, the possible energies which the nucleus can have should be changed slightly and the differences could be ascertained by measurement of the resonance frequency for energy changes of the atom in the field. Again the effect was sought over a period of a day, with some correlation with the direction to the galactic center being expected. The results were negative, and the Yale group set a limit, $\Delta m/m_0 < 10^{-22}$; Drever was able to set a somewhat lower limit, $\Delta m/m_0 < 10^{-23}$.

Coconni and Salpeter, in suggesting that the effect of distant mass on the inertial mass of objects on the earth might be found by such measurements, suggested

6. V. W. Hughes, H. G. Robinson, and V. Beltran-Lopez, *Phys. Rev. Letters* 4 (1960):342.
7. R. W. P. Drever, *Phil. Mag.* 6 (1961):683.

that $\Delta m/m_0$ should be of the order of 3×10^{-10}, whereas experiment indicates that the ratio is less than 10^{-23}. Clearly one must conclude that the search for anisotropy in the inertial mass of atoms and nuclei on the earth because of some interaction with the mass of the galaxy gives negative results. And the first reaction must be that this is a serious blow for Mach's principle and any attempt to account for inertia by an interaction with other mass. Two arguments have been advanced, however, to say that the conclusion is invalid, or at least suspect.

S. T. Epstein[8] pointed out that the calculation of Coconni and Salpeter assumed that all the anisotropy should be seen in the kinetic energy term, and he suggested that may not be true. In every case in which such effects have been sought the total energy of the system can be considered to be a sum of kinetic energy terms and potential energy terms. If we think of the simplest, the oscillating crystal, it seems reasonable to expect that dependence of inertial mass upon orientation would affect the kinetic energy term only, for mass in the form of many crystal atoms is moving and being accelerated as it executes harmonic motion; and the potential energy is the energy of compression of the crystal, specifically energy in the electromagnetic field, as ions are crowded more closely together. Epstein pointed out, however, that one should expect anisotropy in the electromagnetic field also, with the result that both effects need to be computed before one can decide whether the frequency of the crystal should depend upon its orientation relative to the center of the galaxy.

---

8. S. T. Epstein, *Nuovo Cimento* 16 (1960):587.

Similar arguments can be given in each of the other physical systems studied, for both kinetic energy and potential energy terms always are involved in the total energy. Epstein demonstrated that it is not at all inconceivable that the potential energy depends upon the mass anisotropy in such a way as to exactly compensate for the dependence of the kinetic energy, with the result that no measurements of the sort that have been made could demonstrate the effect.

A more general argument was given by Dicke[9], who suggested that this attempt to verify Mach's principle may have been based upon too narrow a view of the principle, and that in fact a proper view may imply that all mass and all fields must respond in the same way to distant mass concentrations, so that all such measurements must give a null result. By this argument, the null result is evidence in support of Mach's principle! The contention is that according to Mach's principle the anisotropy of inertial mass should affect all particles and fields in the same way; consequently, it must be possible at any point to find a coordinate system which is inertial, and in such a system the mass anisotropy cannot be seen. In other words, if Mach's principle is valid, no locally performed experiment can possibly show mass anisotropy. One can conclude then that distant mass affects all local mass and fields in the same way, although that may be a null effect, that is, it may affect them alike by not affecting them at all.

In conclusion, the experiments which have been described in this chapter, culminating in the most sensitive one by Drever, tell us something about the detec-

9. R. H. Dicke, *Phys. Rev. Letters* 7 (1961):359.

tability of mass anisotropy; and in particular they set an extremely low limit on it. What conclusions we may wish to draw from these negative results is not so certain; but in any case if we hope to find experimental verification or refutation of Mach's principle, we must look further, and to a different type of experiment.

5

Mach's Principle and General Relativity

We have considered the attempts to determine whether the inertial mass of bodies depends upon an interaction with the other mass of the universe by looking for effects arising from the fact that the mass of the galaxy is not isotropically distributed about us, and we have found that the result of the search is negative. We have seen also that the failure to find such an effect does not indicate that Mach's principle is invalid if we take a sufficiently broad view of it. Our only hope now of finding experimental indications of the validity of the principle seems to be to look critically at the general theory of relativity to determine whether it is consistent with Mach's principle and whether the theory of relativity is itself supported by experimental evidence. In reality we must consider two theories, the general relativity of Einstein and the scalar-tensor theory of Brans and Dicke. These two theories have somewhat different relations to Mach's principle, and the attempts to decide which of the two is the better description of the world may tell us something about the value of Mach's principle as a part of physical theory.

Discussing the general theory of relativity and Mach's principle, Einstein listed three effects which he would expect the theory to predict if it is consistent with Mach's ideas:

1. The inertia of a body must increase when ponderable masses are piled up in its neighborhood.
2. A body must experience an accelerating force when neighbouring masses are accelerated, and in fact, the force must be in the same direction as that acceleration.
3. A rotating hollow body must generate inside of itself a "Coriolis field," which deflects moving bodies in the sense of the rotation, and a radial centrifugal field as well.[1]

The general theory does predict the second and third effects, although they are very small, but it does not predict the first. Einstein mistakenly thought he had demonstrated that the theory predicts an increase of inertia when ponderable masses are brought near, but it has since been demonstrated that the proof depends upon choice of a particular coordinate system and in fact the theory does not exhibit this effect. General relativity does fail to express Mach's principle in at least one important way that Einstein pointed out: the theory predicts that in a universe empty save for one object, that object should still have inertia.

To make a discussion of the comparison of these two theories (the Einstein theory and the Brans-Dicke theory) intelligible and to explain how either can incorporate Mach's principle, we must gain some understanding of the physical reasoning in the theories; and to this end we now turn to a consideration of the gen-

---

1. A. Einstein, *The meaning of relativity* (Princeton: Princeton Univ. Press, 1923). (Reprinted by permission of the Princeton University Press and the Estate of Albert Einstein.)

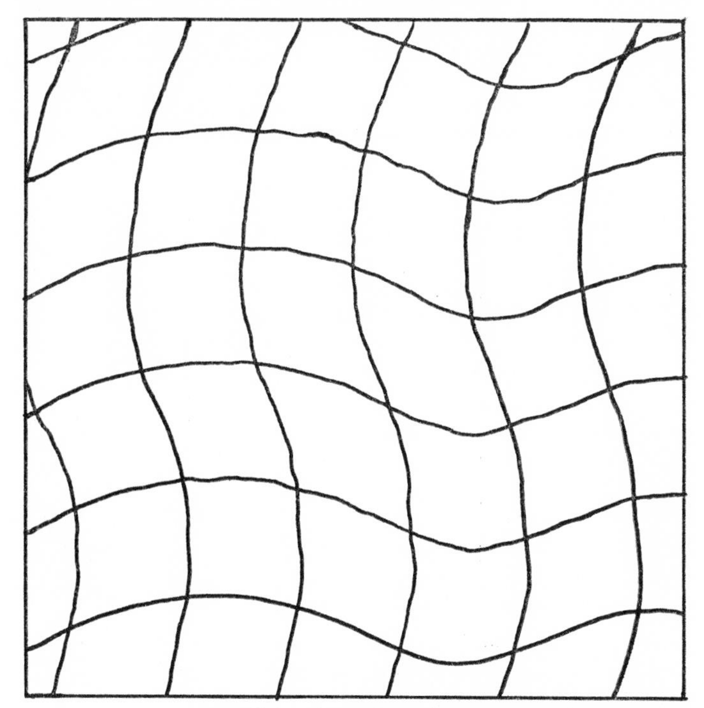

FIG. 5.1. A surface (not necessarily a plane) with an arbitrarily drawn grid of coordinate lines.

eral theory of relativity, stripped of most of the mathematics.

General relativity is an essentially geometric theory, relating points in a four-dimensional space-time. One can visualize a small region in this way: picture a surface covered with a network of lines as shown in Figure 5.1. The lines need not be straight and they need not be perpendicular to one another where they cross. All that is required is that each point on the surface be the intersection of two lines and only two, so that the

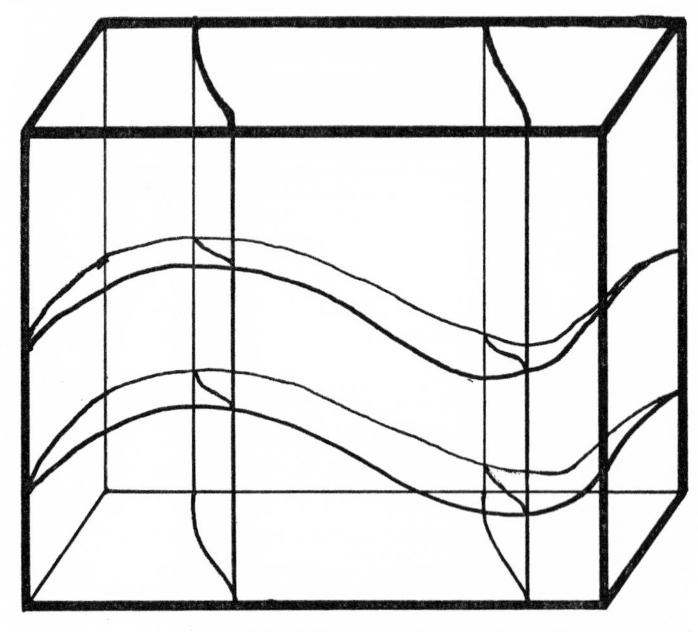

FIG. 5.2. The extension to three dimensions of the arbitrary coordinate grid, with surfaces replacing the lines of Figure 5.1.

position of that point can be given by telling two numbers. Now imagine each of the lines extended upward to form a cylinder, and the surface repeated many times as shown in Figure 5.2, so that any point in a volume of three-dimensional space is the intersection of three surfaces. Any point in the three-dimensional space can now be identified by giving three numbers. Now we imagine at each point a small clock, and we shift our thought from points to "events," an occurrence at a point and at a time. Each event can be identified by giving four numbers, three to tell where and one to tell when it happened. It is not to be inferred that time and distance are equivalent or that one can move backward in time as he can move backward along one of the space directions; but formally time can be treated as a dimension, and it is this four-dimensional space-time that provides the conceptual basis for general relativity.

Not only is the geometry of relativity four-dimensional, but it is also Riemannian. A Riemannian geometry is one in which the distance between adjacent points is defined, so that one can say the square of the distance between the point (x_1, x_2, x_3, x_4) and the infinitesimally distant point $(x_1 + dx_1, x_2 + dx_2, x_3 + dx_3, x_4 + dx_4)$ can be written as

$$ds^2 = \sum_{i=1}^{4} \sum_{j=1}^{4} g_{ij}\, dx_i\, dx_j.$$

Usually the summation signs are omitted, and it is understood that in such an expression when a subscript is repeated it is to be summed. Thus one usually writes only

$$ds^2 = g_{ij}\, dx_i\, dx_j$$

and this stands for a sum of 16 terms of the form

$$ds^2 = g_{11}\, dx_1^2 + g_{12}\, dx_1\, dx_2 + \cdots + g_{44}\, dx_4^2.$$

In general the g_{ij} are not constants, but functions of the x's. For example, in two dimensions on a plane if one uses rectangular coordinates, the distance between two points is given by

$$ds^2 = dx^2 + dy^2.$$

In this case $g_{11} = g_{22} = 1$, and $g_{12} = g_{21} = 0$. On the other hand if one expresses the same distance in terms of polar coordinates, the result is

$$ds^2 = dr^2 + r^2 d\theta^2,$$

from which $g_{11} = 1$, $g_{22} = r^2$, and $g_{12} = g_{21} = 0$.

The reader may be forgiven if his first reaction is to wonder why one bothers to stipulate that the distance between points is defined. Is it not always possible to find the distance between adjacent points? The answer is No. One often uses "spaces" in which distance is not defined. A simple example of such a space which is likely to be familiar to a physics student is the space used in thermodynamics diagrams, in which the axes are p, v, and T. Adjacent points may represent physical systems with slightly different pressures, volumes, and temperatures or they may represent the same system with different parameters; but it makes no sense to ask how far apart the points are. "Distance" has no significance in such a space; hence it is not a Riemannian space.

One of the simplest and most obvious ideas of physics was used by Einstein in the development of both the special and the general theories of relativity, with results that seem surprisingly more sophisticated than the original idea. The idea is that certain aspects of the physical world should be the same regardless of how we choose to decribe them—in technical language, that they should be invariant under changes of coordinates. This was not an original thought with Einstein, but he made particularly effective use of it. The reason for introducing vector analysis into physics is that it permits manipulation of the things which correspond to physically real quantities, the vectors, instead of manipulation of the components which depend upon the particular choice of coordinate system used. In the same way, the distance between two points on this sheet of paper is the same whether one uses rectangular coordinates to express it and writes $ds^2 = dx^2 + dy^2$ or whether he uses polar coordinates and writes $ds^2 = dr^2 + r^2 d\theta^2$. To express it more generally, if we take x_1 and x_2 to represent any coordinates, the distance squared which can be expressed as

$$ds^2 = g_{11}\, dx_1^2 + 2g_{12}\, dx_1\, dx_2 + g_{22}\, dx_2^2$$

must be the same regardless of what coordinate system we may choose to carry out the calculation. As one goes from one coordinate system to another the *dx's* change and the *g*'s change, but *ds* does not. That is what is meant by saying it is invariant. The kinematical equations of special relativity are developed by applying the requirement that the speed of light be an invariant as one changes the relative translational speed of coordinate systems. This is a rather special sort of invariance, for it is applied only to a particular class of coordinate systems; but such surprising results as time dilation still arise directly from the requirement that the mathematical description agree with what seems to be the physical fact, namely that observers moving relative to one another measure the same velocity for light. Another way of saying the same thing is to define a quantity called the "interval" between two events slightly separated in space or in time, or in both, by the expression

$$d\tau^2 = c^2\, dt^2 - dx^2 - dy^2 - dz^2$$

and to require that this be invariant under changes from one coordinate system (x, y, z, t) to another moving relative to it $(x'\ y', z', t')$.

In developing the general theory of relativity, Einstein generalized this idea of the invariance of the interval by proposing that in four-dimensional space-time the distance or interval ds should be invariant under all coordinate transformations; that is, that

$$ds^2 = g_{ij}\ dx_i\ dx_j$$

expressing the square of the interval between two events slightly separated in space or time or both, should be the same regardless of what coordinate system is used.

There are 16 separate values of the g_{ij}, but since $g_{ij} = g_{ji}$, only 10 of them are really different. They can be displayed in an array like this:

$$\begin{matrix} g_{11} & g_{12} & g_{13} & g_{14} \\ g_{12} & g_{22} & g_{23} & g_{24} \\ g_{13} & g_{23} & g_{33} & g_{34} \\ g_{14} & g_{24} & g_{34} & g_{44}. \end{matrix}$$

Because the g's change in a particular way as one changes from one coordinate system to another, this array is called a tensor. It is a mathematical animal having a character of its own, just as a vector is another sort of animal. Whereas a vector in four-dimensional space would have four components, this tensor (of the second order, since it is a two-dimensional array of numbers) has 16 components. It is possible to define tensors of higher rank, and in four-dimensional space the number of components of a third-rank tensor is $4^3 = 64$, of a fourth rank tensor is $4^4 = 256$, and so forth.

One may now ask a very simple question: Are there combinations of the g's or functions of them which are invariant under coordinate changes if ds^2 is invariant? The answer to the simple question can be given simply—it is Yes, but proving this is true and finding the forms of the functions of the g's which are invariant is another matter, and it is anything but simple.

We shall content ourselves with the statement that functions of the derivatives of the *g*'s can be found which are invariant under all coordinate transformations, and that they are related to the curvature. The simplest of these functions is a second-rank tensor, and it is the one used by Einstein; hence it is often called the Einstein tensor.

The notion of curvature is easy to see in one or two dimensions, and one can generalize the formulas to three and four dimensions; but visualizing curvature in more than two dimensions probably is impossible. To understand the idea as clearly as possible, we shall begin with one dimension and work upward. We begin with a line. If it is straight, its curvature is zero, and we may say that it is a segment of a circle whose radius is infinite. If the line is not straight, we pick a point on it and there construct the "osculating" circle, that is, the circle which is tangent to the line at the point and most nearly "fits" the line at the point. The circle has some radius r, and the reciprocal of that radius is defined to be the curvature, or curvature $= 1/r$.

When we deal with a surface, we may pass a plane through the surface to generate a curve of intersection and then determine the curvature of that curve. In general the curvature at a point depends upon the orientation of the intersecting plane. If we find the planes for which the curvature at a point has maximum and minimum values, those planes are perpendicular, and the product of the maximum and minimum curvatures is called the Gaussian curvature; it is a characteristic of the surface itself and can be found without measuring radii outside the surface. The expression from which the Gaussian curvature is computed is the curvature tensor derived from the g_{ij}; it depends upon an involved combination of the second derivatives of the g_{ij} with respect to the variables. The curvature of a plane is zero, of a cylinder made by rolling up a plane, zero, and of a sphere, $1/r^2$. If the surface is shaped like a saddle so that the two osculating circles lie on opposite sides of the surface, the curvature is negative. A quantity closely related to the curvature of a surface is the ratio of circumference to radius of a circle drawn on the surface. On a plane, the ratio is, of course, $2\pi r/r = 2\pi$. On a sphere the ratio is less than 2π. (Visualize a globe and think of the circles made by the latitude lines. If the radius is the distance down from the north pole, the ratio of circumference to radius begins near 2π for very small circles about the pole, is 4 at the equator, and approaches zero as the circles come closer and closer to the south pole.) On a saddle surface, the ratio of circumference to radius of circles (loci of points equidistant from a given point) is greater than 2π, with the excess depending upon the size of the circle.

In three-dimensional space the notion of radius of curvature seems to make little sense, for we cannot visualize it. The discussion of circles can be generalized in a useful way, however. In a three-dimensional space of zero curvature (a flat space) the ratio of the surface area of a sphere to the square of its radius is $4\pi r^2/r^2 = 4\pi$. In a space of positive curvature the ratio is less than 4π, and in a space of negative curvature it is greater than 4π. A similar difference exists for the sum of the interior angles of a triangle. On a plane the sum is 180°, on a surface of positive curvature, such as a sphere, it is greater than 180°, and on a surface of negative curvature it is less than 180°. Probably the first attempt to decide whether the space in which we actually live is Euclidean was made by Karl Friedrich Gauss, who sent beams of light along the sides of a triangle having vertices on hills some distance apart and tried to find a discrepancy between the measured sum of the angles and the 180° predicted by Euclidean geometry. If such a discrepancy exists at all, it is much too small to be measured by this method on the earth; if it could be carried out on the surface of a neutron star the results might show the effect Gauss was seeking.

The simplest function which can be constructed from the g_{ij} and their derivatives which is invariant under change of coordinate systems is really not very simple, for it involves the g_{ij} plus their first and second derivatives. One can call it a single thing only because it is one of the animals called a tensor; and when it is written with subscripts instead of being written out with all components shown, it can be represented by a single symbol. Since this thing is invariant, Einstein argued that it should have physical significance, and he proposed that in empty space it should have the value zero; it is then equivalent to a set of 16 differential equations.

To decide whether assigning the value zero to this function related to the curvature of the four-dimensional space-time has any physical significance, one must decide how the geometrical properties of the space which that implies are related to ordinary physics. In the laboratory one can find instruments to measure the radius of curvature of the sides of a lens, but he will not find an instrument designed to measure the radius of curvature of three-dimensional space, much less of four-dimensional space.

If we set Einstein's invariant function equal to zero and solve the resulting differential equations, we find a set of g_{ij} which are presumed to represent empty space. The property of empty space which we know that might be related to the g_{ij} is that in such space massive bodies move in straight lines at constant speed. Anticipating also the development of the theory for regions of space where matter is present, we know that bodies which are accelerated by gravitational attraction have accelerations which are independent of their masses or compositions. The behavior of the bodies, in other words, is as if it were determined by the space through which the body is passing and were not affected by the nature of the body itself. Both these facts suggest that the motion of bodies either in empty space or in gravitational fields should be determined by the description of the space itself, that is, by the g_{ij}. Einstein was led by considerations of this sort to propose that bodies in empty space or bodies acted on only by gravitational forces move along geodesic paths in space-time.

A geodesic is the "shortest distance" between two points. On a plane the geodesic is a straight line, on the surface of a sphere it is an arc of a great circle, in three-dimensional Euclidean space it is a straight line. Such an explanation seems to offer a way to decide where bodies may go within space-time but to provide no description of motion. Eventually we wish to be able to offer an explanation for gravitation, which acts like a force; hence our description must contain something which plays the role of a force, causing motion. To say that the geodesic path on the earth between Rome and New York passes over Ireland explains why an airplane may take that route but tells us nothing about the forces which cause the airplane to move. How can the idea of geodesics lead to a description of motion and force?

The analogy with lines on a plane and arcs on a sphere is only partially applicable and may be somewhat misleading, for in relativity we are talking about geodesics in space-*time.* When one calculates geodesic paths by well-known mathematical manipulations, the equations which result are differential equations; and if the geometry within which one is working contains time as a coordinate, those differential equations also contain time. In fact, some of the resulting terms are accelerations, and for the case of empty space the requirement that a body move on a geodesic becomes the statement that its acceleration must be zero. This is simply another way of saying it moves on a straight line at constant speed, the fact which we already knew. It appears, therefore, that the idea of geodesic motion provides a link between the geometry of space and real physics, at least for this almost trivial case.

The invariant function of the g_{ij} which has been found is a second-rank tensor. If it is set equal to zero, the result is a description of geometry which is appropriate for empty space. But if the theory is to reproduce approximately the known motion of bodies in gravitational fields, the function must be set equal to something other than zero in regions where mass already exists; the only thing a second-rank tensor can be set equal to is another second-rank tensor. Such a tensor can be constructed from elements such as mass and energy, which we expect to determine the gravitational field; it is called the energy-momentum tensor, having as its elements relativistic mass energy, kinetic energy, and momentum. Einstein equated this tensor to the geometric tensor and then demonstrated that for the case of a weak field the proper choice of an arbitrary constant makes the geodesic motion of masses as described by the relativistic theory identical with the motion of a body moving freely in a gravitational field; this case describes our ordinary experience.

We have now reached the goal; we have a description of geometry which is independent of choice of coordinates and which associates a curvature with each point in space-time. We have equations which relate that curvature to the mass-energy content of the space, so that greater mass means greater curvature of space. And finally we have a method of deciding how a freely moving massive particle (or light beam) will move, for we assume that it follows a geodesic. It is a beautiful and convincing theory, but one must still ask the question which distinguishes physics from mathematics—Does the theory describe the real world? We shall defer consideration of this question for a short time and consider first the relation of the theory of relativity to Mach's principle, and we shall also discuss the ideas of the Brans-Dicke theory.

We return to the three effects Einstein suggested should be tests of the compatibility of Mach's principle and the theory of relativity. The first effect is that the inertial mass of a body should increase if other masses are brought near. Einstein thought that he had demonstrated that general relativity does predict this effect, but in 1962 C. H. Brans[2] demonstrated that the supposed proof depends upon the choice of coordinate sys-

2. C. H. Brans, *Phys. Rev.* 125 (1962):388.

tem and in fact relativity theory does not predict the effect. In oversimplified form the crux of the argument can be given thus: Relativity describes the interaction of masses only through the metric tensor (the g_{ij}), but it is always possible to choose coordinates in a small region of space so that the space appears flat, that is, so that it is the space described by special relativity. In such a space the acceleration produced by a gravitational field is present, but there are no effects which can change the inertial mass of a test particle. Since one can choose a particular coordinate system in which the effect being sought does not appear, it cannot be real.

The second effect suggested by Einstein is that the acceleration of bodies near the test body should seem to produce a force on it. General relativity does predict this effect.

The third effect is that a rotating mass shell should produce inside itself a Coriolis force acting on moving masses and a centrifugal force. This is called the Lense-Thirring effect, named for the men who investigated in detail its prediction by general relativity. The theory does predict this, and it can be seen most clearly by considering a gyroscope. Imagine a large, thick shell of some dense material which is rotating about an axis. Inside the shell is a gyroscope set with its axis perpendicular to the axis of rotation of the shell. If the shell were not rotating, the gyroscope would remain stationary relative to the stars; but because of the rotation of the mass shell the gyro precesses slowly. This effect is predicted because the energy-momentum tensor which is used to determine the g_{ij} which describe directions inside the shell contains terms which arise from the energy and momentum of the rotating shell. The predicted precession rate depends upon the mass of the shell and its angular velocity.

Dicke has raised an interesting question in regard to the Lense-Thirring effect: From the viewpoint of Mach's principle, rotation of the shell or rotation of the gyro or rotation of the matter of the universe outside the shell should all be essentially equivalent. Let us imagine the gyro fixed, with the shell rotating about it and the universe outside rotating slowly (at the rate we formerly described as the precession rate of the gyro). One may now think of the effects of the shell and the outer universe as being in opposition, for the rotations are in the opposite sense. But a strange asymmetry appears, for the formulas for the rotation rate contain the mass of the shell, but not of the universe. In other words, the effect produced by the shell can be changed by changing its mass or its radius, but the theory indicates that the mass of the universe outside or the distance to it could be changed with no effect. Surely a theory which indicates that nothing inside the shell would be changed if half a dozen galaxies were suddenly brought close to the outside must be incomplete. At the very least, it does not express Mach's principle.

In an attempt to incorporate Mach's principle into the theory more satisfactorily, Brans and Dicke[3] in 1961 proposed a modification of general relativity. In essence they made a more complicated interaction between a test mass and the other mass of the universe. Whereas general relativity has a single interaction based upon the connection between the geometry of space "seen" by the test mass and the energy-momentum tensor de-

3. C. H. Brans and R. H. Dicke, *Phys. Rev.* 124 (1961):925.

scribing the other mass, they introduced a scalar function, depending upon the masses contributing to the gravitational field and modifying the geometry. Particles still move along geodesic paths, but the paths are not exactly what they would have been without the action of the scalar field. The effect of the scalar field may be simply described by saying that it causes the gravitational constant *G* to vary, both from place to place and also in time. (An alternative form of the theory makes *G* constant but then permits the masses of particles to vary.)

Since the Brans-Dicke theory was proposed in an attempt to bring Mach's principle into gravitation theory, one must now ask how well it has succeeded. Brans[4] has examined this theory to determine how well it satisfies Einstein's first criterion, that the inertial mass of a body increases if other masses are piled up near it. He concludes that this effect is present in the scalar-tensor theory, whereas it is not present in the original form of relativity. The third of Einstein's criteria is also satisfied better than in the original theory, for in the example discussed—a rotating shell of matter with a gyroscope inside—the scalar-tensor theory predicts about the same precession but further predicts that if the mass of the universe outside is reduced the precession rate will increase, indicating that the "braking" effect of outside mass is decreasing, permitting the gyro to more nearly follow the rotating mass.

In summary, then, one may say that, from the standpoint of incorporation of Mach's principle, the inertial mass of bodies should depend upon an interaction with other mass in the universe, the general

4. C. H. Brans, *Phys. Rev.* 125 (1962):2194.

theory of relativity is only slightly satisfactory, and the Brans-Dicke theory is more satisfactory, although it may not be the last word. If experiment should indicate that it is a better description of nature than the Einstein theory, that fact could be taken as an indication that Mach's principle is an important statement about the world. If the Einstein theory seems to be preferred by nature, the fact would not necessarily mean that Mach's principle is worthless as a guide in the formulation of theory, since the principle could be valid but the Brans-Dicke use of it incorrect; but confidence in Mach's principle as a significant statement about the world would be shaken.

In physics the ultimate test is not the elegance of the theory or its philosophical attraction but its correspondence with the real world, so we now turn to consideration of the experiments which give an indication of the validity of the general theory of relativity and may permit us to decide whether it or the Brans-Dicke theory is preferable. There are three tests which are often referred to as the "classical tests of general relativity," and in addition a fourth test has been proposed and applied. We shall consider them in this order: gravitational redshift, precession of the perihelion of Mercury, the deflection of light, and the time delay in radar signals passing the sun.

The gravitational redshift has been measured with fair precision, and the result agrees with prediction; but unfortunately that fact provides none of the information we want because it can be predicted without invoking either of the relativity theories. The effect is that light coming from a region of strong gravitational field to a region of smaller gravitational field shows a redshift. For example, the wavelength of the yellow light emitted by sodium (the *D* lines) can be measured in the laboratory with very high precision; but if the light emitted by sodium atoms on the surface of the sun is examined with the same care, the wavelength is found to be slightly longer. If the light originates on the surface of a white dwarf star, the shift to the red end of the spectrum is even greater because of the more intense gravitational field on the white dwarf. Because of turbulence in its atmosphere and other complicating effects in stars, these measurements do not have the precision one would like; and the most accurate measurement which has been made of this effect was made in a laboratory by means of the Mössbauer effect. This measurement will be described later.

First we look at the derivation of the redshift to be expected, and we shall find it in two ways. Neither requires any reference to the machinery of general relativity theory.

One of the derivations is based on the principle of equivalence. This principle, which was one of the most important guides to Einstein in developing the general theory of relativity, is the statement that a static gravitational field is equivalent in all respects to an acceleration. For example, in a laboratory on the earth dropped objects fall with an acceleration of 9.8 m/sec^2, a mass of 1 kg stretches a spring balance with a force of 9.8 newtons, and so forth. If one were in a laboratory far out in space, away from gravitational fields, and if the laboratory were accelerating upward at 9.8 m/sec^2, the experimenter inside would find exactly the same behavior for falling objects and spring balances as he found on the surface of the earth; in fact he would not be able to tell from measurements made within the laboratory whether he was motionless in a gravitational field or was being accelerated upward in empty space. Looking at atoms in a strong gravitational field, then, is equivalent to looking at atoms being accelerated away from us; but atoms being accelerated away would exhibit a Doppler shift, and this is exactly the gravitational red-shift.

If the gravitational field is such as to produce an acceleration g and the observer is a distance d from the source so that the light travel time is d/c, the velocity which an observer in an actually accelerated system would gain during the travel of the light is gd/c. First-order Doppler theory predicts a frequency change proportional to this, that is,

$$\frac{\Delta \nu}{\nu} = \frac{v}{c} = \frac{gd}{c^2}.$$

By the principle of equivalence, the same shift would be seen whether it is caused by the relative motion of the source and observer or by a gravitational field.

The second derivation makes use of an even simpler principle—conservation of energy. Imagine a photon which rises a distance d through the gravitational field of the earth, where the acceleration produced by gravity is g. Now if we consider a photon whose energy initially is $h\nu$, we make use of the result of special relativity that the inertial mass associated with this energy is $E/c^2 = h\nu/c^2$, and use the equivalence between inertial mass and passive gravitational mass which is implied by the Eötvös experiment to deduce that the gravitational mass of the photon is also $h\nu/c^2$. We conclude, therefore, that the change in potential energy of the photon as it rises a distance d is $h\nu gd/c^2$ (comparable to the mgd familiar from elementary physics). But by conservation of energy we must conclude that the increased potential energy must be gained at the expense of its intrinsic energy, and thus that the frequency must be reduced, from ν to ν'. Equating the two energy change terms we find

$$h\nu - h\nu' \cong \frac{h\nu}{c^2} gd.$$

Rewritten, this becomes

$$\frac{\nu - \nu'}{\nu} = \frac{\Delta\nu}{\nu} = \frac{gd}{c^2}.$$

This same result can be obtained from the equations of the general theory of relativity, but the equations suggest a slightly different interpretation. The change produced by the gravitational field occurs in the time, so that clocks located in positions of differing gravitational fields run at different rates—those where the field is greatest running at the lowest rate. Since an atom emitting radiation is like a clock, the atoms where the field is greatest emit radiation whose frequency is lower than that emitted where the field is less; the lower frequency is equivalent to a longer wavelength.

Measurements of the gravitational redshift have been made from observations of radiation from the sun and also from white dwarfs, but the most precise have been made in a laboratory on the earth. In Jefferson Physical Laboratory at Harvard, Pound and Rebka[5] and later Pound and Snider[6] measured the gravitational

5. R. V. Pound and G. A. Rebka, Jr., *Phys. Rev. Letters* 4(1960):337.
6. R. V. Pound and J. L. Snider, *Phys. Rev.* 140 (1965):B788.

redshift by means of the Mössbauer effect, as photons emitted by Co^{57} traveled 75 feet up or down a tower to be absorbed by Fe^{57}. Measurements were made with the photons traveling both up and down to eliminate certain geometrical factors. The fractional shift in frequency expected for the two-way trip is

$$\frac{\Delta\nu}{\nu} = 2gd/c^2 = 4.905 \times 10^{-15}.$$

The best measurements of the shift, made by Pound and Snider, gave a value which was 0.9990 ± 0.0076 times this value, and the conclusion of the experimenters is that the result agrees to within 1% with the prediction based on the principle of equivalence.

The first of the three "classical tests of relativity" has thus given a positive result; however, although a negative result would have been cause for dismay, the positive conclusion gives little reason for jubilation. The principle of equivalence is supported, but no conclusions can be drawn about general relativity. Since the prediction of the redshift does not depend upon details of the theory, the Brans-Dicke theory does not make a different prediction; therefore this experimental verification offers no way of deciding between the competing theories.

The second of the well-known tests is the precession of the perihelion of Mercury. The perihelion of a planet's orbit is the point at which it comes nearest the sun; and if one draws a line from the sun to that point, he finds that the line is not stationary in space, but moves (precesses) around the orbit. This precession is primarily the result of the attraction of the other planets, and most of it can be explained by carrying out detailed computations of the effect such attractions are expected to have. In the case of Mercury, however, after all the expected effects of other planets have been considered, the perihelion still has a precession which is unexplained; similar effects, but so much smaller that they are useless as tests of relativity, are present for Venus and for the earth.

For Mercury, when the effects of other planets have been taken into account, an unexplained precession of 43 seconds of arc per century is left. One of the spectacular successes of Einstein's theory, when it was first published, was explanation of this discrepancy, for the general theory of relativity predicts just this amount of rotation more than classical theory does. It would appear that one could hardly ask for better experimental confirmation, for this prediction from relativity theory really does depend upon details of the theory—it cannot be obtained by the sort of arguments we used for the gravitational redshift.

The Brans-Dicke theory predicts a precession about 90% as large as that predicted by the Einstein theory, about 39 seconds of arc per century. Although this seems to be in disastrous conflict with observation, Dicke has pointed out that a possible cause of motion of the perihelion has generally been neglected, and it might be just enough to account for the extra 4 seconds. The effect is that due to oblateness of the sun. If the sun is oblate, that is, if it has an equatorial bulge, the extra attraction of the equatorial region of the sun on Mercury could cause its orbit to shift, so that the perihelion would move approximately 4 seconds of arc per century. If that were the case, then the discrepancy between observation and classically explained effects would be just 39 seconds, which is the amount of rotation the Brans-Dicke theory can explain.

Clearly the question which must be answered is, Is the sun oblate? and Dicke and Goldenberg[7] undertook an answer. The work has been continued by Dicke and other collaborators, and the conclusion which has been reached is that the sun is indeed oblate, by just the amount needed to account for 4″ of the precession of the orbit of Mercury! The work has been done carefully, and errors of a magnitude to invalidate the conclusion about the oblateness of the sun seem unlikely; but it is to be hoped that someone else will repeat this measurement with other equipment.

The idea that the sun might be oblate receives confirmation from an unexpected source, however. The outer parts of the sun rotate with a period of approximately 27 days (the exact period depends upon the latitude, for the outer part of the sun rotates more slowly in the polar regions than at the equator), but one possible explanation for the observed oblateness is that the inner part of the sun rotates much more rapidly, so that centrifugal forces cause an equatorial bulge. If the entire sun once rotated with a period of about 5 days, the magnetic field extending out from the surface and carried out by the solar wind could exert a torque on the surface layers which would have slowed the outer layers to their observed speed. If one examines the rotation rates of stars for which the rate can be observed, he finds that stars which are very hot have high rotation rates; as one moves to cooler and cooler stars, he finds that the rotation rate drops rather abruptly. Hot stars do not have an outflow of particles like the solar wind;

7. R. H. Dicke and H. M. Goldenberg, *Nature* 214 (1967):1294.

hence they may lack a mechanism to slow the outer parts. If one computes the rotation rate that the sun would be expected to show if it were very hot, he finds just the rate that the core should have now to explain the measured oblateness. A consistent picture can be formed of the rotation of stars, then, which is like this: All stars form with high rotation rates; but if a star is cool enough to have a magnetic field carried out into space, its outer layers are slowed. The interior continues to spin at a high speed, however, and in the case of the sun this interior spin is enough to produce oblateness. Undoubtedly the same oblateness would exist in other stars also, but we cannot observe it in any except the sun.

In summary, then, one can say that some form of relativity theory is undoubtedly implied by observations of the precession of the perihelion of Mercury. At the present time, observations of the oblateness of the sun and the calculations based on that oblateness give a slight edge to the Brans-Dicke theory over the Einstein theory, but there are enough uncertainties in the argument and the measurements that this cannot be considered a decisive test.

The third classic test is the bending of light passing the sun. The effect has been observed during eclipses of the sun when stars can be seen whose light passes very near to the edge of the solar disk. Because of the effect of the gravitational field of the sun, the light is bent slightly toward the sun, so that the star appears to be displaced from its normal position. The displacement is measured by comparing two photographs, one taken during eclipse and the other taken from the same place, with the same instruments, six months later when the earth is directly across its orbit from the original position, so that the relative positions of the stars are the same but they can be observed at night.

The first attempts to measure the gravitational bending of light were made in 1919, and numerous attempts have been made since. The experiment is beset with many difficulties which render precision difficult, and the measurements scatter badly. To illustrate the problems, the photograph taken during eclipse conditions is made when the atmosphere is at daytime temperature but perhaps made turbulent by the sudden cooling that accompanies the eclipsing of the sunlight; the second photograph is made at night, when the atmosphere is very likely to be cooler. Since the displacement to be measured is very slight, the variation in atmospheric refraction caused by the temperature variations can introduce a large error.

The bending predicted by the theory of general relativity for starlight just grazing the edge of the sun is 1.75″. One can make a classic calculation of the deflection the sun should cause, assuming that the light consists of photons of mass m (the magnitude of m is not important because it drops out of the equations), and this calculation gives a value just half that predicted by relativity. The Brans-Dicke theory predicts approximately 93% of the value predicted by Einstein's theory. Measurements have given values for the deflection ranging from 1.5″ to almost 3″, with the average lying somewhat above 1.75″.[8] One can perhaps say that the prediction of general relativity is supported by the observations with an uncertainty of 10% to 20%. Clearly

8. R. Adler, M. Bazin, and M. Schiffer, *Introduction to general relativity* (New York: McGraw-Hill, 1965), p. 193.

such observations can provide no assistance in choosing between the Einstein and the Brans-Dicke forms of the theory.

The fourth experimental test of relativity is one which could not have been attempted until the 1960s. When light passes near a massive body such as the sun, not only is its path deflected but its passage is slowed. I. I. Shapiro and collaborators[9] have looked for this effect, using the Haystack antenna at the M.I.T. Lincoln Laboratory and microwaves instead of visible light. The experiment consists of sending pulses of radar-frequency radiation past the sun, so that they are reflected from either Mercury or Venus, and then receiving the reflected signal which has again passed by the sun on the return trip. At the frequency used, the ionized gas around the sun through which the radiation passes has negligible effect on it; and any delay in the return of the pulses beyond what would be expected for straight-line, constant-speed passage is attributed to the general relativistic effect. The delay expected from the general theory of relativity (Einstein form) is approximately 200 microseconds, and the Brans-Dicke theory predicts about 93% this much. The factor relating the two theories is the same as that for deflection of light.

The best results from this experiment reported to date are that the delay is 0.9 ± 0.2 times the value predicted by the Einstein theory. Clearly classic theory is not satisfactory, but it is impossible to choose between the Einstein and the Brans-Dicke theories on the basis of this result. This experiment seems to offer the pos-

9. I. I. Shapiro, G. H. Pettengill, M. E. Ash, M. L. Stone, W. B. Smith, R. P. Ingalls, and R. A. Brockelman, *Phys. Rev. Letters* 20 (1968):1265.

sibility of providing a decisive test by the mid-1970s if budget restrictions do not curtail work on it.

Summarizing the four tests of general relativity, we may say that one of them, the redshift, has been verified to a high degree of precision, but that it tells little if anything about the validity of the theories. The other three indicate that classic theory is inadequate but are not sufficiently precise to permit a choice between the Einstein and the Brans-Dicke theories. There is reason to hope that such a choice will become possible on the basis of improvements in the experiments.

6
Absolute Space

THE LACK OF SUCCESS in devising theories which exorcise completely the demon of absolute space and give unambiguous expression to Mach's principle suggests that one should reconsider these problems with the thought that perhaps the attempt is doomed to failure. Perhaps we live in a world in which the only satisfactory explanation makes use of the concept of absolute space, and perhaps the inertia of bodies is the result of a local interaction with the space in their neighborhood rather than with distant masses. Even though experimental results do not yet prove that the attempt to give precise expression to Mach's principle cannot succeed, honesty must compel us to consider the possibility that the failure thus far reflects a fact that the principle is not true, rather than merely a failure to find a correct expression of it. We shall consider, therefore, the idea of absolute space and inertia which is not Machian from the viewpoint of considering how these concepts might be approached now—not how they were considered 100 or even 300 years ago—so that we shall be somewhat prepared if such a view of the world should be forced upon us. It is possible we shall find that such ideas are more congenial than we had supposed, and we shall come to recoil from them less.

At the time Newton introduced the idea of absolute space and Bishop Berkeley objected to it, and even at the time Mach wrote his criticisms of the concept, space was conceived as something static, and probably something of infinite extent. Newton and Berkeley can have had little idea of the extent of the universe, and their ideas of its age and development were probably influenced more by the first chapter of Genesis than by scientific observations. Mach undoubtedly knew of the work of Herschel indicating that we live in a lens-shaped group of stars, and he doubtless knew of the suggestion of Laplace that some of the nebulous objects seen through telescopes are island universes like our own Milky Way galaxy, but such suggestions were purely speculation; only a few distances to stars were known, and the scale of the universe was completely unknown at Mach's time. Of more importance, however, is the fact that the expansion of the universe was unknown even to Mach, for it was not discovered until 1929. The discovery that the universe is expanding gives to space a dynamic character it did not have before and makes worthwhile a rethinking of the arguments against absolute space to make certain they are still convincing.

Space was a vast and unchanging emptiness into which had been placed stars and planets and other objects of the heavens. Space did not interact with the objects within it, but rather it existed independently of them and would have been just the same had they not been present. Critics of the idea could easily suggest that only the material objects were real and point out that the concept of absolute space had a logical fallacy in that no way could be devised to give a unique choice of the one reference system which coincided with absolute space.

To contrast that view of space with the one consistent with current cosmological thinking, let us describe a possible history of the universe. This description is based upon the Friedman model, which seems to be the most satisfactory model developed to date. In the naive form described here it has some problems, but modifications to eliminate these do not change the essentials.

According to this picture of the universe, it began some ten billion years ago as a very hot, small, and dense aggregate of matter and radiation. This was not merely a hot and compact ball of matter in space, but the curvature of space was such that all space was filled with the very dense matter. The temperature was high enough that the elements as we know them did not exist, but everything was broken down to elementary particles. Perhaps at an earlier time even the particles were broken down and the universe was filled with radiation. For some unknown reason the universe began to expand, and the expansion is still continuing. This means that the distance between any two galaxies which are very far apart is increasing. Out to the distances reached by the 200-inch telescope the rate of expansion seems to be constant and isotropic. Along with the increasing distance between the objects which serve as landmarks in the universe goes a change in the curvature of the universe, so that one does not think of the galaxies spreading out into hitherto empty space. Rather, he thinks of the available space increasing, so that space is always filled with matter, eventually condensed into galaxies, spread reasonably uniformly, with the distances between them increasing constantly.

If one can imagine such a universe in which no turbulence existed and the matter remained entirely quiescent, it would be a space filled with gas or dust, with the particles at rest relative to one another except as the scale of everything increased because of the expansion of space. In our universe, turbulence and the effects of gravitation have superimposed onto this uniform and quiet motion local motions of particles into stars—stars moving within galaxies, and galaxies moving within clusters of galaxies. It is useful, however, for purposes of discussing the universe in its large-scale features, to imagine that we can visualize the dust particles as they would have been if aggregation into stars and galaxies had not occurred. In cosmological discussions an observer riding on these dust particles is given the designation "co-moving observer." Such an observer is co-moving because he is at rest relative to the matter in his vicinity if one neglects purely local turbulent motions. Two co-moving observers who are near together are motionless with respect to one another except for the very slow drifting apart which is the result of the expansion of the universe, and co-moving observers who are separated by a great distance are moving apart with a velocity which is proportional to the distance between them.

This substrate of dust particles or co-moving observers constitutes an absolute space. One can imagine a cube defined by eight particles at the corners, which assumed that form in the earliest phases of the expansion and retains it for all time. The cube becomes larger, for all distances in the universe increase; but relative to other matter the cube does not rotate, and any other piece of matter within it remains at rest relative to the corners of the cube; that is, the corners move away from the object inside at a rate proportional to their distances from it. In imagination, at least, we have given meaning to the idea of absolute space, and now we must ask whether in actuality such a notion has physical content in our universe.

Local aggregations, rotations of galaxies, movements of galaxies within groups, and so forth, make the detection of absolute space by watching eight suitably chosen specks of dust impossible. But the same space can be found operationally by looking at matter so far away that the motion it has as a result of the expansion of the universe is so much greater than the local gyrations that those are negligible. In principle one should be able to look in all directions at galaxies very far distant, determine their distances, compute the Doppler redshift their light should exhibit because of the expansion of the universe, and find a coordinate system which would be equivalent to a co-moving observer. To make the argument simple, suppose that six galaxies —lying ahead, behind, to the right, to the left, above, and below—can be found which are all the same distance from us. They should all show the same redshift; and if they do not, we would interpret that fact as being due to a motion of ourselves relative to the co-moving observer in our vicinity. We could therefore compute our speed relative to the co-moving observer, and we would identify him with absolute space. To the accuracy of our observations the universe seems to be isotropic; hence the redshift of galaxies the same distance from us should be the same regardless of their direction.

Another method of determining our motion relative to a local co-moving observer can be proposed based upon observations of the "fireball radiation." Radiation has been found coming to the earth from outer space which is interpreted to be residual from the very hot and compact phase of the universe. At that time the radiation would have been of short wavelength; but as the universe has expanded, the radiation has been expanded and its wavelength has increased. Now it is observed in the microwave region of the electromagnetic spectrum. This radiation has been bouncing around in the universe for a very long time; and as it reaches us, it brings information about the last matter to scatter it. Of course some may have been scattered very recently and nearby, but most of it has traveled a very long distance indeed since it was last scattered by matter. Thus if variations in the wavelength of the radiation coming from different directions could be found, they could be interpreted as Doppler shifts resulting from our motion relative to the average motion of the last matter to scatter the radiation. But that matter is so far away that its local motion should be negligible compared to the motion resulting from the expansion of the universe; hence if we were moving with the co-moving observer in our neighborhood, we should see completely uniform radiation reaching us, and any deviation from uniformity can tell us our own motion relative to the co-moving observer. One measurement of this sort has been reported,[1] and more are certain to be made.

One must conclude that an operational meaning can be given to absolute space—it is possible to define absolute space in such a way that experiments can be

1. E. K. Conklin, *Nature* 222 (1969):971.

planned to measure our motion relative to it entirely apart from mechanics experiments performed within the laboratory. This is a possibility which probably could not have been foreseen by either Berkeley or Mach, and it undercuts some of their objections to use of the concept of absolute space. The fact that by viewing distant objects one can give some meaning to the idea of absolute space does not mean that matter responds to that space, for example, that a rotating bucket of water "knows" that it is rotating relative to the space around it rather than relative to matter somewhere. This raises a question to which there is still no entirely satisfactory answer.

The general theory of relativity is a field theory, which means that it describes the motion of bodies in terms of an interaction between the bodies and a field existing in their neighborhood. It is distinguished in this way from an action-at-a-distance theory, such as Newton's theory of gravitation. The field is described in terms of the g_{ij}, which means that the properties of space itself are the field quantities; these in turn are determined by the distribution of matter everywhere. In this sense, therefore, the theory describes an interaction between matter and the space local to it. The local properties of space determine the geodesic path and thus dictate the acceleration of freely falling bodies or the paths of light rays. For the particular case of free motion, either uniform motion in the absence of gravitational fields or freely accelerated motion in gravitational fields, the theory ascribes the observed phenomena to an interaction between matter and space, with the space acting almost as a causative agent.

This is not the same thing, however, as explaining

inertia in terms of an interaction with space locally. Suppose, for example, that an electric field acts on a charged body producing acceleration. Could the inertia of the body be explained in terms of its effect on local space? Or in terms of the movement of the point at which it produces a slight curvature of space-time? No such explanation has been developed, but neither has such an explanation been proved impossible. A suggestion of the way such an explanation might be sought is provided by the behavior of the electromagnetic field about a moving charge. As long as the charge is moving at constant speed, the electric field about it can be represented by lines which originate at the charge and extend outward continuously and straight. At some distance from the charge and from its path the field acts as if it anticipates where the charge is going to be in the future and continuously changes its direction, so that at all times it points at the position occupied by the charge at that instant. If the charge is accelerated, the field at a distance from it cannot anticipate that change in its state of motion; and the information that the motion has changed travels at the speed of light out along the field lines as a "kink." This kink carries energy, and part of the work done to produce acceleration goes into the radiation thus sent out by means of the changing electromagnetic field. An accelerated mass may produce gravitational radiation, but this cannot be enough to explain its inertial resistance to acceleration; a naive comparison of the electromagnetic and gravitational problems does not yield the desired explanation for inertia. Whether a more sophisticated exploitation of the analogy can produce an explanation is not known.

7
Questions for the Future

THE MOST OBVIOUS CONCLUSION to be drawn from this account of efforts to find answers to questions about preferred reference frames and about the origin of inertia is that the answers are still lacking. Several facts have been established to a high degree of certainty in the search: the repetition of the Eötvös experiment by Dicke and collaborators has demonstrated to a high degree of accuracy that inertial mass is proportional to gravitational mass; experiments on mass anisotropy have demonstrated satisfactorily that it cannot be detected; the measurement by Pound and collaborators of the gravitational redshift has confirmed theoretical predictions to a reasonable precision; and the many ether-drift experiments provide convincing evidence that uniform velocity relative to a preferred reference system cannot be detected by experiments carried out within a laboratory. On the other hand the experiments which might permit a choice between the Einstein and the Brans-Dicke theories are not sufficiently precise to render a decision. At the same time that no experiment has been devised to give direct evidence of the existence in the world of absolute space or a preferred reference system, it has been realized increasingly that the most satisfactory theory describing space implicitly assumes the existence of a preferred frame.

One interpretation of Mach's principle which has not been mentioned thus far deserves attention. That is the suggestion of J. A. Wheeler that Mach's principle perhaps should be thought of as a sort of boundary condition on the solutions of Einstein's equations. The field equations of general relativity are differential equations and (like all differential equations) their solutions are not complete until boundary conditions or initial conditions are put into the solutions. Many solutions of the equations are possible, and Wheeler has suggested that those inconsistent with Mach's principle should be eliminated from consideration as being nonphysical. Whether solutions can be found which exhibit Machian properties of the sort which Brans and Dicke have tried to incorporate into their theory is unlikely, but a slightly modified Mach's principle could serve as a guide to the acceptability of solutions to the field equations.

It is reasonable to expect that improved measurements, particularly of the time delay of radar signals passing the sun, will indicate that either the Einstein or the Brans-Dicke theory is preferable. No matter what that decision is, however, there will still remain questions about preferred reference systems and Mach's principle. Perhaps these questions will never be answered; perhaps they cannot ever be answered; but perhaps also the answer awaits only a new idea and the result of asking nature a question which has not yet been formulated.

INDEX